TURBULENCE AND SHELL MODELS

Turbulence is a huge subject of ongoing research. This book bridges modern developments in dynamical systems theory and the theory of fully developed turbulence. Many solved and unsolved problems in turbulence have equivalences in simple dynamical models, which are much easier to handle analytically and numerically.

This book gives a modern view of the subject by first giving the essentials of the theory of turbulence before moving on to shell models. These show much of the same complex behavior as fluid turbulence, but are much easier to handle analytically and numerically. Any necessary maths is explained and self-contained, making this book ideal for advanced undergraduates and graduate students, as well as for researchers and professionals, wanting to understand the basics of fully developed turbulence.

PETER D. DITLEVSEN is a Professor at the Niels Bohr Institute, University of Copenhagen. As a theoretical physicist, he researches in the fields of turbulence, statistical physics, dynamical systems and climate dynamics.

TURBULENCE AND SHELL MODELS

PETER D. DITLEVSEN

The Niels Bohr Institute, University of Copenhagen

CAMBRIDGE
UNIVERSITY PRESS

CAMBRIDGE
UNIVERSITY PRESS

University Printing House, Cambridge CB2 8BS, United Kingdom

One Liberty Plaza, 20th Floor, New York, NY 10006, USA

477 Williamstown Road, Port Melbourne, VIC 3207, Australia

314-321, 3rd Floor, Plot 3, Splendor Forum, Jasola District Centre, New Delhi - 110025, India

79 Anson Road, #06-04/06, Singapore 079906

Cambridge University Press is part of the University of Cambridge.

It furthers the University's mission by disseminating knowledge in the pursuit of
education, learning and research at the highest international levels of excellence.

www.cambridge.org
Information on this title: www.cambridge.org/9780521190367

First published 2011

A catalogue record for this publication is available from the British Library

Library of Congress Cataloging in Publication data
Ditlevsen, Peter D.
Turbulence and shell models / Peter D. Ditlevsen.
p. cm.
ISBN 978-0-521-19036-7
1. Turbulence. 2. Fluid dynamics. I. Title.
QA913.D58 2011
532′.0527–dc22   2010033854

ISBN 978-0-521-19036-7 Hardback

Contents

Preface

Fluids have always fascinated scientists and their study goes back at least to the ancient Greeks. Archimedes gave in "On Floating Bodies" (c. 250 BC) a surprisingly accurate account of basic hydrostatics. In the fifteenth century, Leonardo da Vinci was an excellent observer and recorder of natural fluid flows, while Isaac Newton experimented with viscosity of different fluids reported in *Principia Mathematica* (1687); it was his mechanics that formed the basis for describing fluid flow. Daniel Bernoulli established his principle (of energy conservation) in a laminar inviscid flow in *Hydrodynamica* (1738). The mathematics of the governing equations was treated in the late eighteenth century by Euler, Lagrange, Laplace, and other mathematicians. By including viscosity the governing equations were put in their final form by Claude-Louis Navier (1822) and George Gabriel Stokes (1842) in the Navier–Stokes equation. This has been the basis for a vast body of research since then.

The engineering aspects range from understanding drag and lift in connection with design of airplanes, turbines, ships and so on to all kinds of fluid transports and pipeflows. In weather and climate predictions accurate numerical solutions of the governing equations are important. In all specific cases when the Reynolds number is high, turbulence develops and the kinetic energy is transferred to whirls and waves on smaller and smaller scales until eventually it is dissipated by viscosity. This is the energy cascade in turbulence. The difference in size between the scales where kinetic energy is inserted into the flow and the scales where it is dissipated as heat is huge. It ranges from, say, the whole atmosphere of the planet to the sub-millimeter scale where viscosity of the air is important.

This fundamental aspect of turbulence can be illuminated by shell models. Shell models have, through their simplicity, contributed to the understanding of symmetries, scaling and intermittency in turbulent systems. Their relatively low number of degrees of freedom in comparison to high Reynolds number flows has enabled them to bridge the gap between the chaotic dynamics observed in low dimensional systems and turbulence. Their computational affordability and simplicity also make them ideal tools for students entering the field of turbulence, and for researchers to test ideas.

This book gives an introduction to the field of turbulence in the spirit of the Kolmogorov phenomenology represented by the famous "K41" scaling relation. The emphasis on shell models in their own right is that the governing equations for shell models share many aspects and are structurally similar to the Navier–Stokes equation, and they are just so much easier to handle. The book is intended for researchers and professionals who want a fast introduction to the problem of isotropic and homogeneous turbulence in the spirit of dynamical systems theory. It should be accessible for advanced undergraduate and graduate students. Most of the material has been used for teaching the subject at the graduate level. For that, a set of problems can be found at the end of each chapter. There are two types of problem: some address the concepts, the mathematics or completion of the calculations leading to the results in the text. Other problems introduce concepts or phenomena, such as Burgers equation, not treated in the main text. An asterisk * indicates a difficult excercise. Shell models are perfect "lab-systems" for numerical investigations, both for testing new scientific ideas and for students to reproduce theoretical results and to familiarize with concepts like scaling relations, Lyapunov exponents and intermittency. To maintain the flow of the main text, some of the mathematical and technical details are deferred to appendices.

1
Introduction to turbulence

Fully developed turbulence is the notion of the general or universal behavior in any physical situation of a violent fluid flow, be it a dust devil or a cyclone in the atmosphere, the water flow in a white-water river, the rapid mixing of the cream and the coffee when stirring in a coffee cup, or perhaps even the flow in gigantic interstellar gas clouds. It is generally believed that the developments of these different phenomena are describable through the Navier–Stokes equation with suitable initial or boundary conditions. The governing equation has been known for almost two centuries, and a lot of progress has been achieved within practical engineering in fields like aerodynamics, hydrology, and weather forecasting with the ability to perform extensive numerical calculations on computers. However, there are still fundamental questions concerning the nature of fully developed turbulence which have not been answered. This is perhaps the biggest challenge in classical physics. The literature on the subject is vast and very few people, if any, have a full overview of the subject. In the updated version of Monin and Yaglom's classic book the bibliography alone covers more than 60 pages (Monin & Yaglom, 1981).

The phenomenology of turbulence was described by Richardson (1922) and quantified in a scaling theory by Kolmogorov (1941b). This description stands today, and has been shown to be basically correct by numerous experiments and observations. However, there are corrections which are not explainable by the Kolmogorov theory. These corrections are deviations in scaling exponents for the scaling of correlation functions. The Kolmogorov theory is not based on the Navier–Stokes equation, except for one of the very few exact relations, namely the four-fifth law, describing the scaling of a third order correlation function. A final theory explaining the corrections should be based on the Navier–Stokes equation.

Shell models of turbulence were introduced by Obukhov (1971) and Gledzer (1973). They consist of a set of ordinary differential equations structurally similar to the spectral Navier–Stokes equation. These models are much simpler and numerically easier to investigate than the Navier–Stokes equation. For these models a scaling theory identical to the Kolmogorov theory has been developed, and they show the same kind of deviation from the Kolmogorov scaling as real turbulent systems do. Understanding the behavior of shell models in their own right might be a key for understanding the systems governed by the Navier–Stokes equation. The shell models are constructed to obey the same conservation laws and symmetries as the Navier–Stokes equation. As well as energy conservation, the models exhibit, depending on a free parameter of the model, conservation of a second quantity which can be identified with helicity or enstrophy. This second quantity signifies whether the models are 3D turbulence-like where helicity is conserved, or 2D turbulence-like where enstrophy is conserved.

In the case of 3D helical (non-mirror symmetric) turbulent flow, there exists a dual cascade of energy and helicity to small scales. A wave number in the inertial range smaller than the Kolmogorov wave number, where the helicity dissipation becomes important, has been identified. In the case of shell models the flow becomes non-helical from this wave number on, until energy is dissipated around the Kolmogorov wave number. In the case of 2D turbulence a forward cascade of energy to small scales is prohibited altogether by the cascade of enstrophy. On the contrary, the energy is transported upscale in an inverse cascade. In the case of shell models we can investigate under what circumstances this is compatible with equipartitioning of the conserved quantities in a quasi-equilibrium as predicted by equilibrium statistical mechanics.

The corrections to the Kolmogorov scaling expressed through the anomalous scaling exponents can be qualitatively understood as a consequence of intermittency in the energy cascade. By simulation we can make a qualitative link between the multi-fractal cascade models and the shell models. The relative simplicity of the shell models also makes it possible to describe the dynamics in terms of bifurcations, routes to chaos, Lyapunov exponents, and so on using the tools developed in the theory of chaos for low dimensional dynamical systems.

This chapter presents a review of some of the main characteristics and unknowns of turbulence. Turbulence is the chaotic and apparently random flow of a stirred fluid. Fluid flows vary a lot depending on the boundaries containing the flow, stirring, and heating. The flow in the atmosphere of a rotating planet is different from the convection in a pot of boiling water. However, as long as the length scales in the flow are small in comparison to the largest scales, determined by the boundaries, and large in comparison to the molecular mean free path scales, all flows seem to have common characteristics. Turbulence is this common characteristic of the flows.

1.1 The Navier–Stokes equation

Fluid mechanics is the description of fluids on scales large in comparison to the mean free path length of the molecules constituting the fluid. In this limit the fluid is regarded as a continuum characterized completely by a velocity field $u_i(\mathbf{x}, t)$, a temperature field $T(\mathbf{x})$, a pressure field $p(\mathbf{x})$ and a density field $\rho(\mathbf{x})$. At each point x_i the fluid is then fully characterized by the six field variables: three components of velocity, pressure, temperature, and density. In order to determine the evolution of these we need six equations. These are derived from momentum conservation, mass conservation, energy conservation, and the equation of state. In any concrete setting some of the field variables might be approximately constant and the number of equations reduced. When considering fully developed turbulence the fluid is traditionally regarded as incompressible, which is a rather good approximation for water and air. This immediately eliminates the equation specifying density from consideration. When buoyancy can be neglected the temperature variations decouple from the momentum and continuity equations and we are left with a fluid described by the velocity and the pressure field. The dynamics of such a fluid is described by the Navier–Stokes equation (NSE)

$$\partial_t u_i + u_j \partial_j u_i = -\partial_i p + \nu \partial_{jj} u_i + f_i, \tag{1.1}$$

and the continuity equation

$$\partial_i u_i = 0. \tag{1.2}$$

The NSE describes the conservation of momentum. In this book we mainly use the tensor notation: $\partial_i u_j \equiv \partial u_j / \partial x_i$, $\partial_{ij} u_k \equiv \partial^2 u_k / \partial x_i \partial x_j$, etc., and the Einstein convention of summing repeated indices; $\partial_{kk} f \equiv \Delta f$ denotes the Laplacian of f. The equation states that the acceleration of a fluid particle equals the sum of the forces acting on the fluid particle (per unit mass). The left hand side is the material derivative of the velocity field, where the second term is the advection. The first term on the right hand side is the pressure gradient force, the second is the viscous friction (viscosity) and the last term represents all other forces per unit of mass. The last term is, historically speaking, not a part of the NSE but we will specify it whenever convenient.

The continuity equation is the equation for conservation of mass, where in the case of an incompressible fluid the density does not appear. The inverse of the density, which normally appears in front of the gradient of pressure in (1.1) is thus also absorbed in the units of pressure. From these (four) equations, together with boundary and initial conditions, the three components of the fluid velocity u_i and the pressure p can in principle be determined.

However, no general solutions to the NSE are known and a solution can be found only for very simple laminar flows. Pressure can immediately be eliminated from the NSE using the continuity Equation (1.2). Assuming the force to be rotational, $\partial_i f_i = 0$, we obtain a Poisson equation for the pressure by applying the divergence operator to the NSE

$$\partial_{ii} p = -\partial_i u_j \partial_j u_i, \tag{1.3}$$

which we can formally solve by applying the inverse Laplacian

$$p = -\partial_{kk}^{-1}(\partial_i u_j \partial_j u_i). \tag{1.4}$$

We make sense of the inverse of a differential operator when expressing the NSE in terms of Fourier components.

The NSE can be brought to a dimensionless form by defining

$$x = L\tilde{x}, u = U\tilde{u}, t = (L/U)\tilde{t}, \tag{1.5}$$

where L is the length scale of the largest variations in the flow. Note that L would typically be the size of the container or basin for a bounded flow or the size of an obstacle in an extended uniform flow; L is called the integral or outer scale; U is the typical velocity difference at this length scale. We can think of U as the typical velocity when coarse graining the flow at the length scale L. As it is derived from Newton's second law the NSE is Galilean invariant. This means that adding a uniform velocity, say, by moving the frame of reference, does not change the NSE. Thus the overall uniform center of mass velocity is unchanged (in the case that the sum of external forces vanishes) and only velocity differences are important. From L and U we can build a timescale $T = L/U$, which is just the time it takes the fluid at uniform velocity U to travel the distance L. Inserting this into (1.1) and dropping the tilde "~" gives the NSE in dimensionless form:

$$\partial_t u_i + u_j \partial_j u_i = -\partial_i p + Re^{-1} \partial_{jj} u_i + f_i, \tag{1.6}$$

where we have defined the dimensionless Reynolds number

$$Re \equiv \frac{UL}{\nu}, \tag{1.7}$$

and absorbed a factor U^2/L into the forcing term. The pressure gradient term is, as can be seen from (1.4), dimensionally the gradient of a velocity squared, so that it scales with changing units of length and time like the advection term. All terms except for the viscosity are now of order unity. The viscosity is of the order of

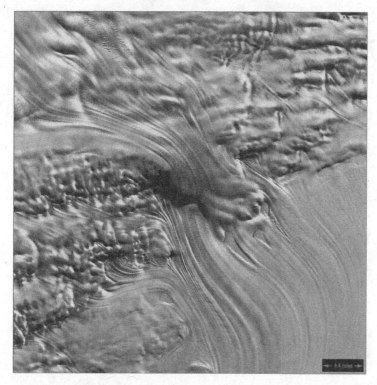

Figure 1.1 Low Reynolds number flow of ice where the viscosity completely dominates the flow. The pheomenon is called an icefall. The fall is about 400 m at the Lambert Glacier, Antarctica (NASA/Landsat).

the inverse of the Reynolds number, so that the Reynolds number is a measure of the relative importance of the viscosity in comparison to the nonlinear terms (the advection and the pressure gradient term) at the length scale L and velocity scale U. The Reynolds number is the fundamental characteristic of any given flow. For a Reynolds number smaller than one the flow will quickly be damped by viscosity, or the viscous term will balance the external forces such as gravity, as is the case in Figure 1.1. The viscosity acts as a smoother of irregularities and has the form of a diffusion term. When the Reynolds number becomes larger the flow will be more and more dominated by the nonlinear terms.

For small Reynolds numbers the flow is smooth and regular. As the Reynolds number is increased the fluid motion in the wake becomes more and more irregular. Increasing Reynolds number flow can be seen as a successive symmetry breaking. For very high Reynolds numbers the regularity of the von Kármán street shown in Figure 1.2 disappears and the flow is completely chaotic and apparently random. This is called *fully developed turbulence*. It characterizes many systems in nature,

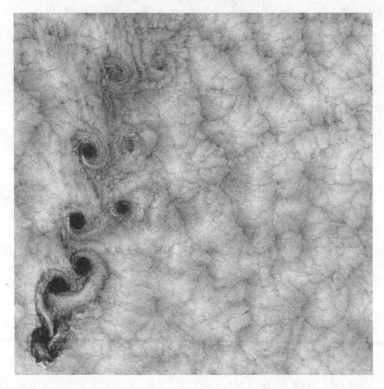

Figure 1.2 Atmospheric flow around Selkirk Island in the southern Pacific Ocean, where the dense cloud cover makes the flow visible. The highest point of the island is about 1.6 km above sea level, obstructing the flow. The phenomenon is called a von Kármán vortex street. (NASA/Landsat)

such as the flow in the atmospheric boundary layer, river flow, the wake after a jet-engine, smoke from a cigar, and many other phenomena. All the richness of the complex behavior of these systems is, we believe, described by the NSE. Direct numerical simulations of the NSE indeed show some of this richness. However, no general theory exists with which we can relate directly the NSE and the rich phenomenology observed in nature and experiments. For high Reynolds number flow there will be a large range of scales where the viscous dissipation is negligible. Assuming either a non-forced decaying motion or forcing restricted to the large scales, motion in this range will be determined by inertia. This is thus called the inertial range. Fully developed turbulence is characterized by a long inertial range. The structure and dynamics of different flows in this range seem in some statistical sense to be alike and one may ask if there is some universality in the behavior of the flows.

The common phenomenology of fully developed turbulence is attributed to Richardson (1922). Richardson describes the flow as consisting of large swirls breaking up into smaller swirls, which again break up into yet smaller swirls until finally the swirls are so small that they are smoothed out, or dissipated, by the viscosity. The energy is inserted into the flow at large scales, it then cascades into smaller and smaller scales until it leaves the flow at the viscous scale. In 1941 this led Kolmogorov to develop a phenomenological theory of turbulence.

1.2 Kolmogorov's 1941 theory (K41)

The paper in which Kolmogorov (1941b) presents the theory is only one half page long and the idea is very simple. It is presented thoroughly by Landau and Lifshitz (1987). Here we will go through it briefly. Kolmogorov imagined a flow initiated by vigorous stirring and then left alone to slowly dampen out by viscosity. This case of unforced flow is today called decaying turbulence. The flow is assumed to be homogeneous (translationally invariant in the mean) and isotropic (rotationally invariant in the mean). The picture we have in mind here is a flow maintained by a force active on large scales of the flow, such that the flow is in a state of statistical equilibrium in the sense that on average the energy input by the force is balanced by the energy dissipated by viscosity (heating the fluid). The state of the flow is then characterized by the mean energy dissipation per unit of mass $\bar{\varepsilon}$ due to viscosity. The velocity characteristic of a given length scale $l \ll L$ is the typical velocity difference $\delta u(l) \equiv |u(r+l) - u(r)|$, where for clarity we suppress all vector indices. This velocity difference is characteristic of the velocity associated with an eddy of size l. The effect of the larger scale flow velocity is merely to move, or sweep, the eddy through the flow as a rigid body. Likewise, if we consider a much smaller eddy within the larger eddy, the effect of the larger eddy on the smaller is the same as the effect of the larger scale flow on the large eddy. Since there is nothing physically significant about a given length scale l in the flow we assume the flow to be self-similar in the sense that when $l_1 < l_2 \ll L$ the velocity differences are related by $\delta u(l_2) = f(l_1/l_2)\delta u(l_1)$, where f is some universal function. This implies that the velocity difference $\delta u(l)$ can only be a function of the scale l and the mean energy dissipation $\bar{\varepsilon}$. From dimensional counting the only possible relationship is

$$\delta u(l) \sim (\bar{\varepsilon} l)^{1/3}, \tag{1.8}$$

where $\sim$ means proportionality. The eddy turnover time is the typical timescale for a fluid parcel propagating across the size of the eddy with the typical velocity associated with that eddy.

We use these kinds of dimensional argument throughout this book. The scaling relation (1.8) is obtained from the fact that only quantities of the same dimension can be compared. So if we want to establish a functional relationship

$$\delta u(l) = \tilde{f}(l, \bar{\varepsilon}), \tag{1.9}$$

the dimension of the right hand side must be the same as the dimension of the left hand side. Furthermore, the numerical value of the quantity on the left hand side cannot depend on change of units of the quantities on the right hand side which leave the units on the left hand side unchanged. If, for example, we measure length in millimeters and time in milliseconds instead of metres and seconds, the numerical value of velocity is unchanged. However, the numerical values of l and $\bar{\varepsilon}$ change. If (1.9) is to hold regardless of the change of units, f can only depend on the combination of l and $\bar{\varepsilon}$ which has the same dimension as the left hand side. The dimensions are $[\delta u] =$ m/s, $[l] =$ m, $[\bar{\varepsilon}] =$ m^2/s^3, so for $[\delta u] = [l]^{\alpha}[\bar{\varepsilon}]^{\beta}$ we get, $\beta = \alpha = 1/3$. From this we get $\delta u(l) = f[(\bar{\varepsilon}l)^{1/3}]$. By changing the units of velocity, say scaling length by a factor λ, we get

$$\lambda \, \delta u(l) = \lambda f[(\bar{\varepsilon}l)^{1/3}] = f[\lambda(\bar{\varepsilon}l)^{1/3}]. \tag{1.10}$$

Thus we see that f must be a linear function and we obtain (1.8).

The relation (1.8) contains all the essentials of the K41 theory. The scale η at which the dissipation becomes important is called the Kolmogorov, or inner scale, in contrast to L, the outer, or integral scale. From (1.1) we can get an estimate of the rate of change of the energy per unit volume due to dissipation at the scale η, $\bar{\varepsilon} \sim \nu u_i \partial_{jj} u_i \sim \nu \delta u(\eta)^2/\eta^2$. Using (1.8) we get

$$\eta \sim (\bar{\varepsilon}/\nu^3)^{-1/4}. \tag{1.11}$$

So keeping the integral length scale velocity and the mean energy dissipation $\bar{\varepsilon}$ fixed, the Kolmogorov scale depends on the Reynolds number as $\eta \sim \mathrm{Re}^{-3/4}$.

The mean of the square of the velocity difference is called the second order structure function $S_2(l)$. The scaling of $S_2(l)$ is obtained by simply squaring (1.8):

$$S_2(l) \equiv \langle \delta u(l)^2 \rangle \sim (\bar{\varepsilon}l)^{2/3}. \tag{1.12}$$

The mean $\langle . \rangle$ denotes an ensemble average, defined as the average over many realizations of the flow with different initial conditions (drawn from some distribution). Assuming ergodicity, this could as well be a temporal average (in a given set of points), or a spatial average in the case of homogeneity. We will freely assume these three to be equal or use either at our convenience without dwelling more on subtleties regarding the assumption of ergodicity or the distribution of initial conditions. A rigorous discussion can be found in Frisch (1995).

1.3 The spectral Navier–Stokes equation

Consider the Fourier transform of the velocity field, Equations (A.3) and (A.4):

$$\mathcal{F}_- : \quad u_i(\mathbf{k}) = \frac{1}{(2\pi)^3} \int e^{-\iota \mathbf{kx}} u_i(\mathbf{x}) d\mathbf{x}, \tag{1.13}$$

$$\mathcal{F}_+ : \quad u_i(\mathbf{x}) = \int e^{\iota \mathbf{kx}} u_i(\mathbf{k}) d\mathbf{k}, \tag{1.14}$$

using the notation introduced in Appendix A.1.

Transforming by $\mathcal{F}_-$, using (A.9)–(A.11), the NSE (1.1), and the Poisson Equation (1.3) give,

$$\partial_t u_i(\mathbf{k}) = -\iota \int u_j(\mathbf{k} - \mathbf{k}') k_j' u_i(\mathbf{k}') d\mathbf{k}'$$

$$- \iota k_i p(\mathbf{k}) - \nu k_j k_j u_i(\mathbf{k}) + f_i(\mathbf{k}), \tag{1.15}$$

and

$$-k_j k_j p(\mathbf{k}) = -\int (k_i - k_i') u_j(\mathbf{k} - \mathbf{k}') k_l' u_m(\mathbf{k}') d\mathbf{k}' \delta_{lj} \delta_{mi}$$

$$= -\int (k_j - k_j') u_l(\mathbf{k} - \mathbf{k}') k_l' u_j(\mathbf{k}') d\mathbf{k}'$$

$$= -\int k_j k_l' u_l(\mathbf{k} - \mathbf{k}') u_j(\mathbf{k}') d\mathbf{k}'. \tag{1.16}$$

It has been noted that incompressibility implies $k_j' u_j(\mathbf{k}') = 0$. Substitution of $p(\mathbf{k})$ from (1.16) into (1.15) gives the spectral NSE

$$\partial_t u_i(\mathbf{k}) = -\iota k_j \int \left(\delta_{il} - \frac{k_i k_l'}{k^2} \right) u_j(\mathbf{k}') u_l(\mathbf{k} - \mathbf{k}') d\mathbf{k}'$$

$$- \nu k^2 u_i(\mathbf{k}) + f_i(\mathbf{k}), \tag{1.17}$$

where $k^2 = k_j k_j$.

If we consider the flow in a box of size L^3 with periodic boundary conditions, the Fourier transform is substituted by a Fourier series and the integral in (1.17) becomes a sum

$$\partial_t u_i(\mathbf{n}) = -\iota\,(2\pi/L)\,n_j \sum_{\mathbf{n}'} \left(\delta_{il} - \frac{n_i n_l'}{n^2}\right) u_j(\mathbf{n}') u_l(\mathbf{n} - \mathbf{n}')$$

$$- vn^2 u_i(\mathbf{n}) + f_i(\mathbf{n}), \tag{1.18}$$

where the wave vectors are $\mathbf{k}(\mathbf{n}) = 2\pi\mathbf{n}/L$, and $n^2 = n_j n_j$. This form of the NSE is the starting point for the shell models. The partial differential equation (PDE) (1.1) has now been substituted by a hierarchy of coupled ordinary differential equations (ODEs). The nonlinear terms are quadratic in the velocities. The interactions are such that only waves with wave vectors adding up to zero are coupled. Such a set of three waves is called a triad. It can be shown by manipulating indices in (1.17) and (1.18) that the inviscid energy conservation fulfilled by the NSE is a detailed energy balance, so that energy is exchanged within each triad. The algebra involved in proving this and many other relations of the NSE is much simpler, but completely similar in the case of shell models. We thus for transparency do many of the calculations for the case of shell models.

1.4 The spectral energy density

The second order structure function (1.12) is related to the spectral energy density through a Fourier transform, as we now show. The energy density per unit of mass of the flow can be expressed in spectral form by use of Parseval's identity (A.7):

$$E = \frac{1}{2} \int \mathbf{u}(\mathbf{x})^2 d\mathbf{x} = \frac{1}{2}(2\pi)^3 \int_0^\infty u_i(\mathbf{k}) u_i(\mathbf{k})^* d\mathbf{k}$$

$$= \frac{1}{2}(2\pi)^3 4\pi \int_0^\infty k^2 |\mathbf{u}(k)|^2 dk \equiv \int E(k) dk, \tag{1.19}$$

where we have assumed the flow to be isotropic, $u_i(\mathbf{k}) = u_i(k)$, and performed the integration over the sphere. We have absorbed the unit of length into the spatial variable $d\mathbf{x} \to d\mathbf{x}/L^3$, where L is the linear size of the integration box. Thus we define the spectral energy density as

$$E(k) = 2\pi(2\pi)^3 k^2 |\mathbf{u}(k)|^2. \tag{1.20}$$

The Fourier transform of the velocity is expressed in terms of the second order structure function

$$S_2(\mathbf{l}) = \langle (\delta \mathbf{u}(\mathbf{l}))^2 \rangle = \int [\mathbf{u}(\mathbf{l} + \mathbf{x}) - \mathbf{u}(\mathbf{x})]^2 d\mathbf{x}$$

$$= 2 \int [\mathbf{u}(\mathbf{x})^2 - \mathbf{u}(\mathbf{l} + \mathbf{x})\mathbf{u}(\mathbf{x})] d\mathbf{x}. \tag{1.21}$$

By using (1.19), (1.20), (1.21), and $\int e^{\iota \mathbf{kx}} d\mathbf{x} = (2\pi)^3 \delta(\mathbf{k})$ (A.2), we obtain

$$|\mathbf{u}(k)|^2 = \frac{1}{(2\pi)^6} \int \int e^{\iota \mathbf{k(x-y)}} u_i(\mathbf{x}) u_i(\mathbf{y}) d\mathbf{x} d\mathbf{y}$$

$$= \frac{1}{(2\pi)^6} \int e^{\iota \mathbf{kl}} \left[\int u_i(\mathbf{l}+\mathbf{y}) u_i(\mathbf{y}) d\mathbf{y} \right] d\mathbf{l}$$

$$= \frac{1}{(2\pi)^6} \int e^{\iota \mathbf{kl}} [\langle \mathbf{u}^2 \rangle - S_2(l)/2] d\mathbf{l}$$

$$= \frac{1}{(2\pi)^3} \langle \mathbf{u}^2 \rangle \delta(\mathbf{k})$$

$$- \frac{1}{(2\pi)^6} \int_0^\infty \int_0^{2\pi} \int_{-\pi/2}^{\pi/2} e^{\iota kl \cos\theta} \frac{S_2(l)}{2} l^2 \sin\theta d\phi d\theta dl$$

$$= \frac{1}{(2\pi)^3} \langle \mathbf{u}^2 \rangle \delta(\mathbf{k}) - \frac{1}{(2\pi)^6} 2\pi \int_0^\infty \frac{1}{2} \left[-\frac{e^{\iota klx}}{\iota kl} \right]_{x=-1}^{x=1} l^2 S_2(l) dl$$

$$= \frac{1}{(2\pi)^3} \langle \mathbf{u}^2 \rangle \delta(\mathbf{k}) + \frac{1}{(2\pi)^5} \int_0^\infty \sin(kl) \left(\frac{l}{k}\right) S_2(l) dl, \tag{1.22}$$

where we have used homogeneity in the substitution $\mathbf{l} = \mathbf{x} - \mathbf{y}$ and isotropy in performing the integration over the sphere. The first term is the kinetic energy of the motion of the center of mass of the fluid. This can in most cases be eliminated by a simple Galilean transformation to the center of mass frame. Finally we get

$$E(k) = \frac{1}{2\pi} k^{-1} \int_0^\infty x \sin x S_2(x/k) dx, \tag{1.23}$$

with the substitution $x = kl$; this is called the Wiener–Khinchin formula. Inserting the scaling relation (1.12) for the second order structure function we get

$$E(k) \sim \bar{\varepsilon}^{2/3} k^{-5/3}. \tag{1.24}$$

This is the celebrated Kolmogorov scaling of the energy spectrum. This relation could as well be obtained by the same dimensional counting as was used for deriving the scaling relation for the velocity increments. The scaling according to (1.24) in the inertial range of the energy spectrum for developed 3D turbulence has been demonstrated in many experiments and seen in many observations. Figure 1.3 shows

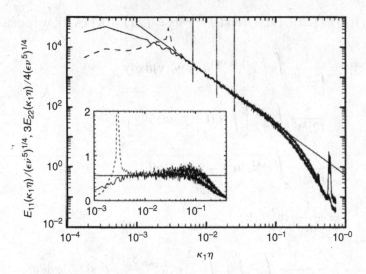

Figure 1.3 The energy spectrum measured using the Taylor hypothesis in a wind tunnel experiment on flow past a cylinder. The two spectra are one dimensional longitudinal spectra E_{11} along the direction of the mean flow (full curve) and E_{22} across the direction of the mean flow (dashed curve). For isotropic turbulence these are simply connected. The straight line is the Kolmogorov scaling $E \sim k^{-5/3}$. The compensated spectrum $E/k^{-5/3}$ is shown in the insert. The peak in E_{22} is due to the periodic passing of Kármán vortices (Kang & Meneveau, 2001)

an example from the atmospheric boundary layer where the Reynolds number is of the order 10^5. In most experiments the velocity is measured at a single point using a hot-wire probe, where the cooling of the wire is proportional to the wind speed. The velocity field is then constructed assuming that the field is frozenly swept past the probe with some mean or large scale velocity $\mathbf{U}$. The velocity field is then reconstructed as $v(x,t) = v(0, t - x/U)$, corresponding to a 1D cut through the velocity field in the direction of $\mathbf{U}$. The validity of this procedure is assumed in the Taylor hypothesis.

1.5 The spectral energy flux

The energy cascade in the Richardson picture is expressed quantitatively through the spectral flux of energy due to the nonlinear terms in the NSE. It is defined as

$$\Pi(k) = \frac{d}{dt}\Big|_{n.l.} \int_0^k E(k')dk', \qquad (1.25)$$

where the derivative denotes the rate-of-change due to the nonlinear terms. From (1.20) and the spectral NSE (1.17) the flux can be expressed explicitly in terms of the spectral velocities:

$$\Pi(k) = \int_{|\mathbf{k}'|<k} (2\pi)^4 k^2 u_i(\mathbf{k}')^* \frac{\mathrm{d}}{\mathrm{d}t}|_{\mathrm{n.l.}} u_i(\mathbf{k}')\mathrm{d}\mathbf{k}' + \mathrm{c.c.} \tag{1.26}$$

$$= \int_{|\mathbf{k}'|<k} (2\pi)^4 k^2 u_i(\mathbf{k}')^*$$

$$\times \left[-\iota k_j' \int \left(\delta_{il} + \frac{k_i' k_l'}{k'^2} \right) u_j(\mathbf{k}'') u_l(\mathbf{k}' - \mathbf{k}'')\mathrm{d}\mathbf{k}'' \right] \mathrm{d}\mathbf{k}' + \mathrm{c.c.},$$

where c.c. denotes complex conjugate. In the case of shell models, this energy flux, which we shall return to later, has a much simpler form.

1.6 The closure problem

The statistics of a turbulent flow is characterized by the correlations between the velocity vectors in different points. In tensor notation we can define a two point correlation function as

$$c_{ij}(\mathbf{x}, \mathbf{x}') = \langle u_i(\mathbf{x}) u_j(\mathbf{x}') \rangle, \tag{1.27}$$

where the bracket denotes an ensemble average. In homogeneous flow the translational invariance implies that this correlation function depends only on the distance vector $\mathbf{x} - \mathbf{x}'$ between the two points. Furthermore, in isotropic flow the rotational invariance implies that the correlation function depends only on the distance $|\mathbf{x} - \mathbf{x}'|$. The Galilean invariance of the Navier–Stokes equation implies that if a constant velocity U_i were added to all velocities in the flow this would again be a solution. However, the correlation function would change to $\langle (u_i + U_i)(u_j' + U_j) \rangle = c_{ij} + U_i U_j + \langle u_i \rangle U_j + U_i \langle u_j' \rangle = c_{ij} + U_i U_j$. The two last terms vanish if the flow is isotropic where any odd order one point mean must be zero. We have introduced the shorthand notation $u_i' = u_i(\mathbf{x}')$. If instead we define correlations between velocity differences as

$$s_{ij} = \langle (u_i - u_i')(u_j - u_j') \rangle, \tag{1.28}$$

we obtain an expression which is independent of the mean velocity of the flow. Furthermore, for $l = |\mathbf{x} - \mathbf{x}'| < L$ the influence of eddies of scales larger than l is merely to sweep the flow at the two points in a rigid motion. This sweeping motion would be present in (1.27) but not in (1.28). In general we would expect the small scale structure of the flow to be better represented by (1.28), which are called structure functions, than by the correlation functions (1.27). The structure functions can trivially be expressed in terms of the correlation functions by expanding the parenthesis

$$s_{ij} = \langle u_i u_j \rangle + \langle u_i' u_j' \rangle - \langle u_i u_j' \rangle - \langle u_i' u_j \rangle = \frac{2}{3} \langle u^2 \rangle \delta_{ij} - 2c_{ij}, \qquad (1.29)$$

where homogeneity and isotropy have been assumed. In order to obtain an expression for the correlation functions, we observe that in the case of decaying turbulence, the correlation will in general be time dependent and we can express the time dependence using the Navier–Stokes equation

$$\partial_t \langle u_i u_j' \rangle = \langle (\partial_t u_i) u_j' \rangle + \langle u_i \partial_t u_j' \rangle$$

$$= \langle (-u_k \partial_k u_i - \partial_i p + \nu \partial_{kk} u_i) u_j' \rangle$$

$$+ \langle u_i (-u_k' \partial_k' u_j' - \partial_j' p' + \nu \partial_{kk}' u_j') \rangle. \qquad (1.30)$$

The derivative ∂' denotes derivative with respect to $\mathbf{x}'$. This means that we can express the evolution of the second order correlation function in terms of third order correlation functions involving velocities and derivatives of velocities. The pressure, using the incompressibility condition, can be expressed by solving the Poisson Equation (1.3) involving a term which is quadratic in the velocities. The dissipation terms are linear and can, by isotropy, be expressed as $\nu \partial_{kk} \langle u_i u_j' \rangle$. This set of equations is not closed in the sense that the number of unknowns is larger than the number of equations. In order to close the set of equations we need equations for the time evolution of the third order correlation functions. These are readily obtained by the same procedure, taking the derivative with respect to time of the third order correlators and applying the Navier–Stokes equation for each term. This will yield an equation involving fourth order correlation functions from the nonlinear terms in the Navier–Stokes equation, and thus needs a specification for those in order to be closed. We finally end up with an infinite hierarchy of equations. This is the closure problem of turbulence. A whole branch of turbulence research deals with development of suitable approximations for closing the equations. The most commonly used closure is to assume Gaussian statistics for the (even orders of) velocity differences from which the higher order correlation functions can be expressed by the pair correlation functions through Wick's theorem. We shall not stroll further along this path here. A recent account and references for various closures can be found in Lesieur (1997).

1.7 The four-fifth law

A special situation occurs for a third order structure function which can be associated with energy conservation. The spectral flux of energy can be expressed in terms of a third order structure function. This can be verified by dimensional counting.

The relation was first derived by von Kármán and Howarth (1938). Since energy is conserved by the nonlinear terms in the Navier–Stokes equation, the mean energy flux must be equal to the mean energy dissipation $\bar{\varepsilon}$. This leads to an exact scaling relation for the third order structure function in the inertial range where energy dissipation can be neglected in comparison to the nonlinear flux of energy. This scaling relation was derived by Kolmogorov (1941a) and is called the four-fifth law. The reason for the name will be apparent shortly. The four-fifth law is derived in great detail in Frisch (1995). A completely analogous relation can be derived for the shell models where the algebra is much simpler and transparent. The following is a short derivation based on Landau and Lifshitz (1987).

Consider a homogeneous and isotropic turbulent fluid, brought about by vigorous stirring and then left alone. The motion of this fluid is governed by the Navier–Stokes equation with no forcing term, so it will eventually have lost all its energy to heat by viscous dissipation and be at rest. This scenario is called decaying turbulence. The mean energy dissipation is (minus) the mean rate of change of kinetic energy, which by interchanging the time derivative and the averaging is given as

$$\bar{\varepsilon} = -\frac{1}{2}\partial_t \langle u^2 \rangle. \tag{1.31}$$

In order to utilize the isotropy we will now orient the first axis of the coordinate system along the distance vector $\mathbf{l} = \mathbf{x} - \mathbf{x}'$ connecting two points. This coordinate is called the longitudinal direction and denoted by the suffix l (which should not be confused with the length of the distance vector). The two perpendicular transversal directions are then by isotropy indistinguishable and denoted by the suffix t. From the $i = j = 1$ component of (1.29) the right hand side can be expressed in terms of the longitudinal components of a two point structure function and a two point correlation function

$$\bar{\varepsilon} = -\frac{3}{4}\partial_t(s_{\mathrm{ll}} + 2c_{\mathrm{ll}}). \tag{1.32}$$

Terms like c_{tl} and s_{tl} vanish because of isotropy. The terms c_{tt} and s_{tt} are related to the longitudinal components, but we shall not need that here. From the Navier–Stokes equation we have the expression (1.30) for the rate of change of c_{ij}

$$\partial_t c_{ij} = -\partial_k(c_{ki,j} + c_{kj,i}) + 2\nu\partial_{kk}c_{ij}, \tag{1.33}$$

where we have defined

$$c_{ki,j} = \langle u_k u_i u_j' \rangle. \tag{1.34}$$

To arrive at (1.33) from (1.30) requires several steps. Firstly, we can interchange $\mathbf{x} \leftrightarrow \mathbf{x}'$ and simultaneously substitute ∂ for ∂' by a simple rotation using isotropy. Secondly, moving the derivative ∂_k outside the averaging is done using incompressibility; $\partial_k u_k = 0$. Thirdly, the terms involving pressure vanish. Again using incompressibility, we have

$$\partial_i \langle u_i p' \rangle = 0. \tag{1.35}$$

The vector $\langle u_i p' \rangle$ can, because of isotropy, only depend linearly on the unit vector $\hat{l}_i = l_i/l, l = |\mathbf{l}|$, in the direction $\mathbf{l} = \mathbf{x}' - \mathbf{x}$. Thus $\langle u_i p' \rangle = f(l)\hat{l}_i$. The general solution to (1.35) of this form is $\langle u_i p' \rangle = (c/l^2)\hat{l}_i$, with c constant. This is only finite for $l \to 0$ when $c = 0$ and we have $\langle u_i p' \rangle = 0$.

To proceed from (1.33) we use isotropy to obtain an expression for the third order correlation function $c_{ki,j}$. This is a third order tensor which is symmetric in the two first suffices. The most general form of this tensor, only depending on $\hat{l}_i$, and the unit tensor δ_{ij}, is

$$c_{ki,j} = A(l)\delta_{ik}\hat{l}_j + B(l)(\delta_{kj}\hat{l}_i + \delta_{ij}\hat{l}_k) + C(l)\hat{l}_k\hat{l}_i\hat{l}_j. \tag{1.36}$$

The prefactor of the two terms in parenthesis must be the same $B(l)$ due to the symmetry between k and i. This excludes the antisymmetric tensor of rank three ϵ_{ijk} as well. We can derive relations between the prefactors $A(l), B(l)$, and $C(l)$ by using incompressibility. Taking the derivative of $c_{ki,j}$ with respect to the second coordinate, we get from (1.34)

$$\partial'_j c_{ki,j} = \langle u_k u_i \partial'_j u'_j \rangle = 0. \tag{1.37}$$

Using the expression (1.36) for the correlation function we get

$$\begin{aligned}
\partial'_j c_{ki,j} = {} & A'\partial_j l \delta_{ki}\hat{l}_j + A\delta_{ki}\partial_j \hat{l}_j \\
& + B'\partial_j l(\delta_{kj}\hat{l}_i + \delta_{ij}\hat{l}_k) + B(\delta_{kj}\partial_j \hat{l}_i + \delta_{ij}\partial_j \hat{l}_k) \\
& + C'\partial_j l\hat{l}_k\hat{l}_i\hat{l}_j + C(\hat{l}_i\hat{l}_j\partial_j \hat{l}_k + \hat{l}_k\hat{l}_j\partial_j \hat{l}_i + \hat{l}_k\hat{l}_i\partial_j \hat{l}_j).
\end{aligned} \tag{1.38}$$

On the right hand side $'$ denotes differentiation with respect to the argument. By straightforward differentiation we have $\partial_j l = \partial_j \sqrt{l_k l_k} = \hat{l}_j$, $\partial_j \hat{l}_i = (\delta_{ij} - \hat{l}_i\hat{l}_j)/l$, and $\partial_j \hat{l}_j = 2/l$, so the expression above becomes

$$\begin{aligned}
\partial'_j c_{ki,j} = {} & A'\delta_{ki} + 2A\delta_{ki}/l + 2B'\hat{l}_k\hat{l}_i + 2B(\delta_{ki} - \hat{l}_k\hat{l}_i)/l \\
& + C'\hat{l}_k\hat{l}_i + 2C\hat{l}_k\hat{l}_i/l.
\end{aligned} \tag{1.39}$$

From (1.37) we get the two equations

$$A' + 2(A+B)/l = 0,$$
$$(3A' + 2B' + C') + (6A + 4B + 2C)/l = 0, \tag{1.40}$$

where the first equation comes from the term proportional to δ_{ki} and the second equation from taking the trace. The second equation can, after multiplication by l^2, be rewritten as $[l^2(3A+2B+C)]' = 0$, from which it follows that

$$3A + 2B + C = c/l^2, \tag{1.41}$$

where c is a constant of integration. For $l = 0$ the correlation function must vanish as any odd order one point correlation function due to isotropy. Thus we have $c = 0$ and we finally get

$$B = -A - A'l/2,$$
$$C = A'l - A, \tag{1.42}$$

leaving us with only one prefactor $A(l)$, so that

$$c_{ki,j} = A\delta_{ki}\hat{l}_j - (A + A'l/2)(\delta_{kj}\hat{l}_i + \delta_{ij}\hat{l}_k) + (A'l - A)\hat{l}_k\hat{l}_i\hat{l}_j. \tag{1.43}$$

Finally, A can be expressed in terms of the third order structure function

$$s_{kij} = \langle(u_k - u_k')(u_i - u_i')(u_j - u_j')\rangle = 2(c_{ki,j} + c_{jk,i} + c_{ij,k}). \tag{1.44}$$

The last identity follows trivially from expanding the parenthesis and using homogeneity and isotropy. From (1.43) we get $c_{ll,1} = -2A$ and from (1.44) we get

$$A = -s_{lll}/12. \tag{1.45}$$

With (1.44) we can now calculate the first term on the right hand side of (1.33), which by a trivial but lengthy calculation gives

$$-\partial_k c_{ki,j} = \frac{1}{l}(-2A + 2A'l + A''l^2/2)\hat{l}_i\hat{l}_j$$
$$+ \frac{1}{l}(-2A - 3A'l - A''l^2/2)\delta_{ij}. \tag{1.46}$$

Combining this with (1.32), the purely longitudinal component of (1.33), and (1.45) we get

$$-\frac{2}{3}\bar{\varepsilon} - \frac{1}{2}\partial_t s_{ll} = \partial_t c_{ll} = \frac{1}{6l}(4s_{lll} + ls_{lll}') + 2\nu\partial_{kk}\left(\frac{1}{3}\langle u^2\rangle - \frac{1}{2}s_{ll}\right). \tag{1.47}$$

The second term on the left hand side can be neglected because it is small in comparison to the first term on scales $l \ll L$. This is because in a decaying turbulent flow the timescale over which the flow can be regarded as (semi-)steady state is of the order of the large scale turnover time L/U. The viscous term is, with use of (1.11), of the order $\nu\delta u(l)/l^2 \sim \bar{\varepsilon}(\eta/l)^{4/3}$. In the inertial range $l \gg \eta$ we can thus neglect the viscous term. The first term on the right hand side can be rewritten as $(l^4 s_{III})'/(6l^4)$, so rearranging the terms we get

$$-\frac{4}{5}(\bar{\varepsilon}l^5)' = (l^4 s_{III})',$$
(1.48)

which upon integration, using $s_{III}(0) = 0$, gives

$$s_{III} = -\frac{4}{5}\bar{\varepsilon}l.$$
(1.49)

This is the four-fifth law which is a cornerstone of fluid mechanics. The scaling relation is in agreement with the phenomenology described in the K41 theory, but the four-fifth law is, in contrast to the K41 phenomenology, exact in the proper limit and among the few exact results derived from the Navier–Stokes equation.

1.8 Self-similarity of the energy spectrum

If a fluid is stirred and then left alone the fluid motion will slowly decrease through viscous transformation of kinetic energy into heat. This situation is termed decaying turbulence. Decaying turbulence is governed by the unforced NSE, which is Equation (1.1) with $f_i = 0$. Decaying turbulence is seen to be invariant under the simultaneous scalings

$$\mathbf{x} \to l\mathbf{x}, \quad t \to l^{1-h}t, \quad \mathbf{u} \to l^h\mathbf{u},$$

$$(p/\rho) \to l^{2h}(p/\rho), \quad v \to l^{1+h}v,$$
(1.50)

where l and h are arbitrary parameters. This is the scaling symmetry of the NSE. It implies that if all variables are scaled according to (1.50) and $\mathbf{u}(\mathbf{x}, t)$ is a solution to the NSE, so is $l^h\mathbf{u}(l\mathbf{x}, l^{1-h}t)$. The scaling relation is just the trivial statement that the equation is independent of the choice of units. However, there are measurable physical consequences due to the chaotic nature of the NSE (Olesen, 1997; Ditlevsen *et al.*, 2004). Consider the energy density (1.19),

$$E = \int_0^\infty E(k, v)dk,$$
(1.51)

where using (1.20) and (1.22) we have

$$E(k, t, v) = \frac{k^2}{4\pi^2}\int e^{\iota \mathbf{k}\mathbf{y}}\langle\mathbf{u}(\mathbf{x}+\mathbf{y}, t)\mathbf{u}(\mathbf{x}, t)\rangle d\mathbf{y}.$$
(1.52)

Here, and in the following, we assume convergence of the relevant integrals. At small scales, large k in (1.51), this is ensured by the presence of the viscosity,

while for the large scales (infrared convergence), this must of course be validated
a posteriori. Assuming homogeneity, the integral in (1.52) is independent of **x** and,
assuming isotropy, the energy density depends only on the modulus $k = |\mathbf{k}|$ of the
wave vector as indicated. The brackets $\langle\rangle$ denote the average with respect to the
ensemble of initial field configurations. Now we can scale all variables according
to (1.50) and obtain the scaling for the energy spectrum

$$E(k/l, l^{1-h}t, l^{1+h}\nu) = l^{1+2h}E(k, t, \nu). \tag{1.53}$$

This scaling relation is not completely trivial since an ensemble mean is involved in
the definition. The ensemble is taken from some distribution of initial conditions.
The distribution of initial fields need not fulfil the scaling relation (1.53). This
implies that if the initial distribution is not "forgotten," as is the case for integrable
governing equations, we should not expect (1.53) to hold. This is the case for
the Burgers equation (She *et al.*, 1992; Vergassola *et al.*, 1994; Gurbatov *et al.*,
1997). The Burgers equation is the NSE without the pressure term, $\partial_t u_i + u_j\partial_j u_i = \nu\partial_{jj}u_i$, corresponding to an infinitely compressible fluid. The Burgers equation is
integrable, so that all statistics, which are highly nontrivial, depend on the ensemble
of initial conditions. The NSE behaves differently, since the chaotic mixing will
make the statistics independent of initial condition, and the scaling relation (1.53)
could be expected for some value of h specified by the initial ensemble.

The scaling relation (1.53) can be simplified further by introducing a fixed unit
of time $t_0 \equiv l^{1-h}t$, such that we can express the energy spectrum at time t in terms
of the energy spectrum at time t_0, which is then a function only of the two variables,
k and ν. Using $l = [t_0/t]^{\frac{1}{1-h}}$, (1.53) becomes

$$E\left(k[t/t_0]^{\frac{1}{1-h}}, t_0, [t/t_0]^{\frac{1+h}{1-h}}\nu\right) = (t/t_0)^{\frac{1+2h}{1-h}}E(k, t, \nu),$$

which implies that

$$E(k, t, \nu) = k^{-1-2h}\left(k[t/t_0]^{\frac{1}{1-h}}\right)^{1+2h}E\left(k[t/t_0]^{\frac{1}{1-h}}, [t/t_0]^{\frac{1+h}{1-h}}\nu\right)$$

$$\equiv k^{-1-2h}\psi\left(k[t/t_0]^{\frac{1}{1-h}}, [t/t_0]^{\frac{1+h}{1-h}}\nu\right). \tag{1.54}$$

Here we have introduced a scaling function, $\psi(x, y) = x^{1+2h}E(x, t_0, y)$ of two
variables. Defining $q = -1 - 2h$ we get

$$E(k, t, \nu) = k^q\psi\left(k^{\frac{3+q}{2}}t, \nu t^{-\frac{1-q}{3+q}}\right). \tag{1.55}$$

The actual form of the function $\psi(x, \nu)$ cannot be found without reference to the
NSE and the initial ensemble. In general, ψ will not behave as a power function

in its arguments. In the limit of small v, we can assume the existence of an inertial range in which the scaling function becomes independent of v and the Kolmogorov spectrum would appear for large k. In this case the scaling function is a power function and we have

$$\psi(x, v) = \psi(x) \sim x^{-2\frac{5+3q}{3(3+q)}} = k^{-5/3} k^{-q} t^{-2\frac{5+3q}{3(3+q)}}, \tag{1.56}$$

where we have used $x = k^{\frac{3+q}{2}} t$.

A scaling relation for the total energy $E(t, v)$ (denoted by its arguments) is obtained by integrating the energy spectrum. Using (1.54) we obtain

$$
\begin{aligned}
E(t, v) &= \int_0^\infty E(k, t, v) \mathrm{d}k \\
&= \int_0^\infty k^q \psi\left(k^{\frac{3+q}{2}} t, v t^{-\frac{1-q}{3+q}}\right) \mathrm{d}k \\
&= (t/t_0)^{-2\frac{1+q}{3+q}} \int_0^\infty \tilde{k}^q \psi\left(\tilde{k}^{\frac{3+q}{2}} t_0, v[t/t_0]^{-\frac{1-q}{3+q}} t_0^{-\frac{1-q}{3+q}}\right) \mathrm{d}\tilde{k} \\
&= (t/t_0)^{-2\frac{1+q}{3+q}} E\left(t_0, v[t/t_0]^{-\frac{1-q}{3+q}}\right), \tag{1.57}
\end{aligned}
$$

by a change of integration variable $\tilde{k} = k \, (t/t_0)^{\frac{2}{3+q}}$. This is valid for all $q > -1$. However, if the dependence on v is treated as negligible, the scaling symmetry is broken in the sense that the parameter q in principle can be determined experimentally from the decay of the total energy.

From the classical K41 (Kolmogorov, 1941c) dimensional counting we have $E \sim k^{-5/3} \epsilon^{2/3}$, where $\epsilon = \mathrm{d}E/\mathrm{d}t$ is the mean energy dissipation. Comparing with (1.56) we have $\mathrm{d}E/\mathrm{d}t \sim t^{-(5+3q)/(3+q)}$, which after integration gives the result obtained in (1.57), ignoring the v-dependence.

These scaling relations can be compared with the experimental measurements of the energy spectrum in decaying turbulence by Comte-Bellot and Corrsin (1971). The energy spectrum was measured in a wind tunnel experiment at different distances from the turbulence generating grid; see Figure 1.4 (top). The scaling parameter q is determined from the data by using (1.57). It is found that $E(t) \sim t^{-1.25}$ for this experiment and it is thus within the range of most experimental findings (Mohamed & LaRue, 1990). The center panels show this scaling behavior and from (1.57) we obtain $q = 2.33$. The bottom panels show the scaling function $\psi(k^{(3+q)/2} t) = E(k, t, v)/k^q$ obtained from (1.55) as a function of $(k^{(3+q)/2} t)^{2/(3+q)}$. The collapse of the different curves to one curve is seen to be perfect. The energy spectra show an inertial range of less than one decade, and smaller for the 1 inch (25.4 mm) grid experiment. So even in this case of relatively low Reynolds number

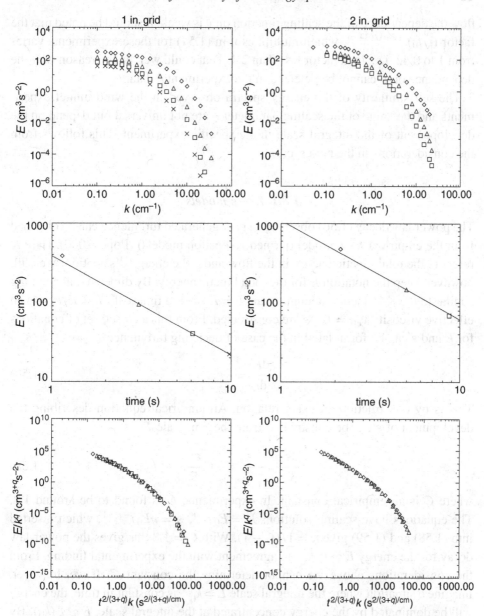

Figure 1.4 Decay of grid turbulence generated in a wind tunnel (Comte Bellot & Corrsin, 1971). Top panels show energy spectra in two experiments with grid spacing of 1 inch and 2 inches (25.4 and 50.8 mm), respectively. The kinematic viscosity corresponds to that of air, $\nu = 1.5 \times 10^{-5}$ m^2/s. The spectra are measured at different distances from the grid, which by Taylor's hypothesis are converted into times using the relation $t = D/U_0$, where D is the distance from the mesh and U_0 is the wind velocity upstream from the grid. The times in the panel are (top left panel): $t = 1.14$ s ($\diamond$), $t = 3.05$ s ($\triangle$), $t = 6.1$ s ($\square$), $t = 9.8$ s ($\times$), and (top right panel): $t = 1.07$ s ($\diamond$), $t = 2.49$ s ($\triangle$), and $t = 4.34$ s ($\square$). Middle panels show the total energy $E(t)$ as a function of time t with the symbols as in the top panels. The straight lines are $E \sim t^{-2(1+q)/(3+q)}$ with $q = 2.33$. Bottom panels show the scaling function $\psi(k^{(3+q)/2}t) = E(k,t,\nu)/k^q$ as a function of $kt^{2/(3+q)}$ with $q = 2.33$.

flow the dependence of the scaling function on ν is weak. It should be noted that the factor $(t/t_0)^{(1-q)/(3+q)}$, which multiplies ν in (1.54) for these experiments varies from 1 to 0.42, i.e., by a factor less than 2.4. That could well be the reason why the dependence on ν cannot be detected in the experimental data.

The self-similarity of the energy spectra observed in the wind tunnel experiments and the value of the scaling parameter q are not universal but depend on the developement of the integral scale in the specific experiment. This follows from the considerations in the next section.

1.8.1 *K* − *ε* models

The power law decay (1.56) observed in grid-generated turbulence can be obtained from the empirical $K-\varepsilon$ model (termed K–epsilon models) (Pope, 2000). Here K refers to the total kinetic energy in the flow and ε the energy dissipation. We will, however, keep the notation E for the total kinetic energy. By dimensional counting similar to the K41 theory, a length scale $L = E^{3/2}/\varepsilon$, a timescale $T = E/\varepsilon$, and an effective viscosity $\nu_{\text{eff}} = E^2/\varepsilon$ are constructed. From this a closed set of equations for E and ε can be formulated in the case of decaying turbulence

$$\frac{dE}{dt} = -\varepsilon. \tag{1.58}$$

This is by construction an exact relation. An empirical equation describing the development of ε can be constructed from the timescale T:

$$\frac{d\varepsilon}{dt} \approx -\frac{\varepsilon}{T} \approx -C\frac{\varepsilon^2}{E}, \tag{1.59}$$

where C is an empirical constant. In experiments, C is found to be around 1.8. The equations have scaling solutions $E = E_0 t^{-n}$, $\varepsilon = nE_0 t^{-n-1}$, which inserted into (1.58) and (1.59) give $n = 1/(C-1)$. With $C = 1.8$ this gives the power-law decay for the energy $E \sim t^{-1.25}$ in agreement with the experimental finding. From the considerations above we cannot determine the constant C. If we, however, imagine a situation where the integral scale $L = k_0^{-1}$ is constant in time, the energy will be dominated by the energy concentrated at the integral scale, $E \sim E(k_0)$. By dimensional counting and assuming a constant energy flux from (1.24) we get

$$\varepsilon \sim E(k_0)^{3/2} k_0^{5/2} \sim E^{3/2}. \tag{1.60}$$

Inserting this into (1.58) and solving for E gives $E \sim t^{-2}$. This is a faster decay rate than the one observed in most wind tunnel experiments. The reason is that in the experiments the integral scale just behind the mesh will be of the order of the grid spacing, while further downstream the integral scale will be determined by the size of the tunnel. This results in an inverse energy cascade to larger scales, slowing the dissipation rate at the small scales. The self-similar decay found in the previous section should thus be taken as an observation rather than a rigorous derivation since the dependence of the energy spectrum on the infrared cutoff (the integral scale) was neglected.

1.9 The dissipative anomaly

The four-fifth law was derived using energy conservation and the NSE. An important consequence of the evolution of the energy can be seen from an analysis of the energy equation

$$D_t \int \frac{1}{2} u_i u_i \mathrm{d}\mathbf{x} = -\nu \int \omega^2 \mathrm{d}\mathbf{x} + F, \qquad (1.61)$$

derived as (A.28) in Appendix A.4. We have here introduced the squared length $\omega^2 = \omega_i \omega_i$ of the vorticity, $\omega_i = \epsilon_{ijl} \partial_j u_l$ of the velocity field $\mathbf{u}(\mathbf{x})$. The energy equation is obtained from the NSE by integration of the kinetic energy $u_i u_i / 2$ over the volume of the flow under the assumption that the boundary conditions are periodic or that the velocities vanish at the boundaries. The pressure term vanishes due to incompressibility. The term F is the rate at which energy is pumped into the flow at large scales.

In a steady state the time mean of the two terms on the right side of (1.61) must be in balance. However, in the case of a vanishing viscosity, $\nu \to 0$, the energy must still remain constant under steady state flow. Consequently

$$\int \omega^2 \mathrm{d}\mathbf{x} \to \infty \text{ as } \nu \to 0. \qquad (1.62)$$

This integral of the squared vorticity is called the *enstrophy*. The divergence of the integral implies that the velocity gradients in the flow become unlimited as $\mathrm{Re} \to \infty$. This phenomenon is called the *dissipative anomaly* and it implies that the energy cascade to the small length scales must be accompanied by a production of squared vorticity.

1.10 The vorticity equation

By applying the curl operator to the NSE as shown in (A.29) we obtain the vorticity equation

$$\partial_t \omega_i + u_j \partial_j \omega_i = \omega_j \partial_j u_i + \nu \partial_{jj} \omega_i + \epsilon_{ijl} \partial_j f_l, \qquad (1.63)$$

The left side is the material derivative of the vorticity, while the last two terms on the right side are dissipation and a stirring force. The first term on the right side appears from the advection term in the NSE, and it is the nonlinear term responsible for the production of vorticity at the small length scales. It is called the stretching and bending term because it can be split as shown in (A.32) into two terms representing vortex stretching and vortex bending, respectively. The first term signifies the creation of vorticity by stretching vortices along the axis of rotation. This is the effect a ballet dancer uses when pulling the arms towards the body to increase the spin in a pirouette. The second term is the creation of vorticity from the bending of vortices caused by velocity shear along the axis of rotation. This is observed in the winding of tornadoes. From the dissipative anomaly and the vorticity equation we expect the energy dissipation and small scale vortices to be intimately linked in fully developed turbulence. The pressure term does not appear in the vorticity equation, since the curl of a gradient trivially vanishes. However, there is no free lunch, solving the Poisson equation (1.3) for the pressure is substituted by solving the equation $\omega_i = \epsilon_{ijk} \partial_j u_k$ for u_k, still with u_k fulfilling the incompressibility constraint.

1.11 Intermittency in turbulence

The four-fifth law establishes an exact scaling relation for a third order structure function. For other orders, the K41 scaling theory predicts the scaling

$$S_p(l) = \langle \delta u(l)^p \rangle \sim l^{p/3}, \tag{1.64}$$

which is the only one that can be obtained by dimensional counting. This would imply that $\langle \delta u(l)^{2q} \rangle \sim \langle \delta u(l)^2 \rangle^q$, as is the case if $\delta u(l)$ is Gaussian. By a field being Gaussian we mean that it has a Gaussian distribution. If the field is not Gaussian, we should not expect (1.64) to hold in general. From the four-fifth law we know that the longitudinal component $\delta u(l)_{\|}$ must have a skewed distribution, since the odd ordered structure functions would vanish for a symmetric distribution. High Reynolds number flow is observed to behave strongly intermittently. Calm periods are interrupted by sudden bursts of energy at small scales which are then effectively dissipated.

There is now conclusive experimental evidence that corrections to the scaling law (1.64) are present in the inertial range of fully developed turbulence. The structure functions still scale with length, but the scaling exponents $\zeta(p)$ defined from

$$S_p(l) \sim l^{\zeta(p)} \tag{1.65}$$

are different from $p/3$. Figure 1.5 shows the scaling exponent $\zeta(p)$ as a function of p for turbulence in a wind tunnel (Anselmet *et al.*, 1984). The straight line is the K41 prediction. The function $\zeta(p)$ is called the anomalous scaling exponents,

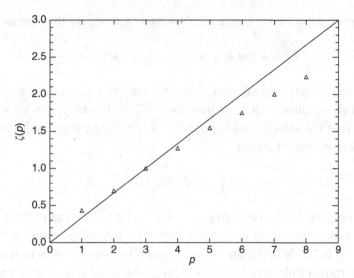

Figure 1.5 The anomalous scaling exponents derived from structure functions measured in a wind tunnel experiment (Anselmet *et al.*, 1984).

anomalous referring to the inhomogeneous nature of the intermittent flow. The bursts of strong energy dissipation are associated with small scale structures of high vorticity. Whether these structures are vortex sheets, vortex tubes, or more complicated filament structures is not completely understood. A phenomenological description of the connection between these structures and the anomalous scaling is given by the fractal or multiplicative models which we describe briefly later. Calculating the anomalous scaling exponents from the NSE is a major challenge in which there has been little success despite a rather large effort in recent years. Shell models show intermittency corrections to the K41-like scaling as well, and for shell models it is relatively easy, at least qualitatively, to understand how the intermittency emerges. Even though shell models cannot account for the spatial structures in fully developed turbulent flow, the intermittent behavior of these models has been a major motivation for studying them.

1.12 Finite time singularities

The dissipative anomaly implies that if the flow is initially a smooth field the velocity gradients will grow in time. It is, today, an open question whether the gradients will blow up in finite time or if there are solutions to the Navier–Stokes equation for arbitrarily high Reynolds numbers which will stay smooth for all times. If the solutions to the NSE are irregular, we must define the differential operators in terms of distributions. Such singular solutions $u_i(\mathbf{x}, t)$ satisfy the integral equation

$$\int (u_i \partial_t + u_i u_j \partial_j + p \partial_i - \nu u_i \partial_{jj} - f_i) \phi_i \mathrm{d}\mathbf{x} \mathrm{d}t = 0, \tag{1.66}$$

where $\phi_i(\mathbf{x}, t)$ is any smooth function with compact support, a so-called test function. This equation is derived from the NSE by, formally, moving all the differentiations of the velocity field $u_i(\mathbf{x}, t)$ by integration by parts to the smooth test field. Correspondingly we have

$$\int u_i \partial_i \psi \mathrm{d}\mathbf{x} \mathrm{d}t = 0 \tag{1.67}$$

for the incompressibility. Here $\psi(\mathbf{x}, t)$ is any scalar test function. Such solutions are called weak solutions of the NSE (Leray, 1934). A naïve argument for a finite time blow-up in the NSE equation can be given in terms of eddy turnover times. The eddy turnover time τ_l at scale l is from dimensional counting proportional to $\overline{\epsilon}^{-1/3} l^{2/3}$, see (1.12), taking the time for an energy burst to be transferred from a scale l to, say, $l/2$ to be proportional to τ_l. This gives a convergent quotient sum and thus a finite time for the energy transfer to infinitely small scales. The question of finite time blow-up in the NSE has been posed as one of only seven "Hilbert problems" for the twenty-first century, with an award for its solution of 1 million dollars by the Clay Mathematics Institute (2000).

1.13 Problems

1.1 **Shear flow.** This problem concerns one of the simplest flows possible. The form of the flow is to be determined as the solution of the Navier–Stokes equation (NSE) with specific boundary conditions. We write the NSE and the condition of incompressibility (continuity or mass conservation):

$$\partial_t u_i + u_j \partial_j u_i = -\partial_i p + \nu \partial_{jj} u_i,$$
$$\partial_i u_i = 0.$$

The flow is constrained by two infinite xz-planes placed at $y = 0$ and $y = D$. See Figure 1.6. The bottom plate is at rest while the top plate moves in the x-direction with constant velocity U_0. This differential motion forces the flow, which is expressed through what is called the *no slip boundary conditions*. This means that the flow is "attached" to the surfaces in such a way that it has the same velocity as the boundary. This is a fundamental property of fluids. The boundary conditions are thus:

$$u_i(x, 0, z) = 0$$

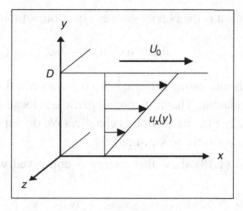

Figure 1.6 Constrained flow.

and
$$u_i(x,D,z) = (U_0,0,0).$$

We now assume the following simple form of the flow:

$$u_i(x,y,z) = (u_x(y),0,0).$$

For this flow write up the NSE (three equations), and check that the flow fulfils the incompressibility condition.
Finally, determine $u_x(y)$.

1.2 **Flow between plates.** With the same geometry as in the previous problem, both plates are at rest ($U_0 = 0$) and the flow is now forced by a pressure gradient $-\Delta p/L$ in the x-direction.
Using the no slip boundary conditions, again determine $u_x(y)$.
A little more advanced, but more relevant for real situations: Consider flow through a pipe in the x direction. The pipe has a circular profile with radius r. The flow is in the x direction with $u_x(y^2 + z^2)$. The boundary condition is $u_x(r^2) = 0$. Calculate u_x by solving the NSE. The resulting velocity profile is called a Poiseuille solution.
Calculate an expression for the total flow (m³/s) as a function of the radius of the pipe, the pressure gradient, the length of the pipe, and the viscosity.

1.3 **Arnold–Beltrami–Childress flow** is defined as,

$$\begin{pmatrix} u \\ v \\ w \end{pmatrix} = \begin{pmatrix} A\sin z + C\cos y \\ B\sin x + A\cos z \\ C\sin y + B\cos x \end{pmatrix}. \tag{1.68}$$

Determine the pressure field in the ABC flow. Then show that the ABC flow is a static solution to the Euler equation.

The Euler equation is the Navier–Stokes equation without viscosity,

$$\partial_t \mathbf{u} + \mathbf{u} \cdot \nabla \mathbf{u} = -\nabla p. \tag{1.69}$$

The ABC flow is interesting because it is a regular static flow in which particle trajectories are chaotic. That is, if a small particle at location $\mathbf{x}$ flows with fluid such that $d\mathbf{x}/dt = \mathbf{u}(\mathbf{x})$ its motion is chaotic. (We do not show that here, but you can do that yourself on a computer.)

1.4 Using Equation (1.18) show that energy is conserved within each triad of waves.

1.5 Derive Equation (1.24) by dimensional analysis.

1.6 **Burgers equation,**

$$\partial_t u + u \partial_x u = \nu \partial_{xx} u, \tag{1.70}$$

is a one-dimensional model of turbulence ($u = u(x,t)$). The inviscid case ($\nu = 0$) is just an advection equation in one dimension, $Du/Dt = 0$.
We now (in the inviscid case) follow a fluid particle initially $t = 0$ at $x(0)$. The motion of the particle is governed by

$$\frac{dx}{dt} = u(x(t), t).$$

From Burgers equation, find the position $x(t)$ of the particle.
The initial velocity field in the interval $0 < x < L$ is $u(x,0) = U \sin 2\pi x/L$. Draw the velocity field at times $t = 0, T/2, T$, where $T = L/U$. What happens for $t > T$?
Define $s(x(t),t) = \partial_x u(x(t),t)$ (the slope). Derive an equation for Ds/Dt by using Burgers equation. Consider this as a differential equation for $s(t)$ (following a fluid particle at $x(t)$), and solve the equation by integration, using the initial condition $s(0) = s(x(0),0) = -S_0$. When does the slope become infinite?

1.7 ***Gaussian closure of Burgers equation.** Derive for Burgers equation the dynamical equations to fourth order for the ensemble averages, $\langle u \rangle$. A Gaussian random variable is defined as a stochastic variable with a Gaussian distribution. Assume that u is a Gaussian random variable and apply the Gaussian closure of the system of equations. Derive an equation for $\langle u^2 \rangle$. The Gaussian closure is (from Wick's theorem):

$$\langle u^4 \rangle = 3\langle u^2 \rangle^2. \tag{1.71}$$

Show this.

1.8 **The Navier–Stokes equation** for an incompressible fluid,

$$\partial_t \mathbf{u} = -\mathbf{u} \cdot \nabla \mathbf{u} - \nabla p + \nu \nabla^2 \mathbf{u} + \mathbf{f},$$
$$\nabla \cdot \mathbf{u} = 0, \tag{1.72}$$

is considered in a periodic box. The spectral representation of the velocity field is then

$$\mathbf{u}(\mathbf{x}) = \sum_{\mathbf{k}} \mathbf{u_k} e^{\imath \mathbf{k} \cdot \mathbf{x}}. \tag{1.73}$$

The velocity field is a real field, so we have $\mathbf{u_k^*} = \mathbf{u_{-k}}$.
Show that the incompressibility implies $\mathbf{u_k} \cdot \mathbf{k} = 0$ for all k.
Express the advection term in spectral representation.
Using incompressibility we get a Poisson equation for the pressure field from taking the divergence of (1.72),

$$-\Delta p = \nabla \cdot (\mathbf{u} \cdot \nabla) \mathbf{u}. \tag{1.74}$$

Use the spectral representation for the advection term to solve this equation for p.
Finally, derive the spectral form of the Navier–Stokes equation (1.18).

1.9 **Solving the Poisson equation for pressure.** The pressure is obtained from the NSE and incompressibility condition by applying the divergence operator to the NSE. The formal solution is

$$p = \Delta^{-1} (\partial_i \partial_j u_i u_j).$$

For a periodic domain the velocity and the pressure can be expressed as a Fourier series:

$$\mathbf{u}(\mathbf{x}) = \sum_{\mathbf{k}} \mathbf{u_k} e^{\imath \mathbf{k} \cdot \mathbf{x}}$$

and

$$p(\mathbf{x}) = \sum_{\mathbf{k}} p_{\mathbf{k}} e^{\imath \mathbf{k} \cdot \mathbf{x}}.$$

Solve the Poisson equation for the pressure by finding an expression for $p_{\mathbf{k}}$.

2

2D turbulence and the atmosphere

2.1 2D turbulence

If flow is constrained to two dimensions, as is the case for some stratified fluids, it is fundamentally different from flow in three dimensions. The free atmosphere on Earth is the prime example of this, thus we describe the flow of the atmosphere in some detail in this chapter.

If we consider the vorticity Equation (1.63) it contains the stretching and bending term which ensures the production of enstrophy $Z = \int \omega^2 d\mathbf{x}$ necessary for dissipation to remain constant in the limit of vanishing viscosity. This term will trivially vanish in two dimensions since the vorticity is perpendicular to the plane of flow. In 2D, the vorticity can be defined as a scalar $\omega = \epsilon_{ij}\partial_i u_j$, where $\epsilon_{11} = \epsilon_{22} = 0$ and $\epsilon_{12} = -\epsilon_{21} = 1$, and the vorticity equation reduces to the scalar equation

$$D_t\omega = \partial_t\omega + u_j\partial_j\omega = -\nu\partial_{jj}\omega + \epsilon_{ij}\partial_i f_j. \tag{2.1}$$

This means that the vorticity acts as a passively advected quantity and the mean square vorticity, the enstrophy, is an inviscid invariant of the flow along with the energy. If we consider an unforced flow at high Reynolds number, then the enstrophy must grow in order to dissipate the energy at small scales. This is, however, not possible since the enstrophy can only decrease from its initial value due to dissipation. Thus a cascade of energy to small scales is impossible in 2D turbulence. What happens then? The energy and enstrophy equations are

$$D_t \int \frac{u_i u_i}{2} d\mathbf{x} = -\nu \int \omega_i \omega_i d\mathbf{x}, \tag{2.2}$$

$$D_t \int \frac{\omega_i \omega_i}{2} d\mathbf{x} = -\nu \int \partial_i \omega \, \partial_i \omega d\mathbf{x}, \tag{2.3}$$

(see (A.52) and (A.43)). The corresponding spectral representations are

$$D_t \int E(\mathbf{k})d\mathbf{k} = -2\nu \int k^2 E(\mathbf{k})d\mathbf{k}, \tag{2.4}$$

$$D_t \int k^2 E(\mathbf{k})d\mathbf{k} = -2\nu \int k^4 E(\mathbf{k})d\mathbf{k}, \tag{2.5}$$

respectively, where $E(\mathbf{k}) = u_i(\mathbf{k})u_i^*(\mathbf{k})/2$ (see (A.54) and (A.48)). These two equations are called the spectral energy equation and the spectral enstrophy equation, respectively.

The integrand on the left side of (2.5) is the spectral enstrophy density $Z(\mathbf{k}) = k^2 E(\mathbf{k})$. The ratio between dissipation of energy and dissipation of enstrophy at wave vector $\mathbf{k}$ is seen to be k^{-2}. Since the dissipation takes place at the Kolmogorov scale η, we have $\int k^2 E(k)dk \sim \eta^{-2}E(\eta^{-1})$. The energy dissipation can be estimated as $\bar{\varepsilon} = \bar{z}\eta^2$ where $\bar{z}$ is the mean enstrophy dissipation. The mean enstrophy dissipation must balance the mean enstrophy input and the energy dissipation must consequently vanish for high Reynolds number flow ($\eta \to 0$). This led Kraichnan (1967) to propose a completely different scenario from the 3D case for cascade in 2D turbulence. The conjecture is that enstrophy and not energy cascades to the small scales. The energy on the contrary cascades to the large scales. So in 2D turbulence there are two inertial ranges, one for the forward cascade of enstrophy and one for the inverse cascade of energy. In the inertial range for enstrophy cascade we can repeat the dimensional scaling arguments from K41 for the enstrophy. The mean dissipation of enstrophy has the dimension $[\bar{z}] = \text{time}^{-3}$. Thus, if the scalar enstrophy spectrum (see (A.50) and (A.49))

$$Z(k) = k^2 E(k) = k^3 \int_0^{2\pi} E(k \cos\theta, k \sin\theta)d\theta \tag{2.6}$$

depends solely on the mean enstrophy dissipation $\bar{z}$ and the scale k^{-1}, we get from dimensional counting, $Z(k) \sim \bar{z}^{2/3} k^{-1}$, and using $Z(k) = k^2 E(k)$ we obtain the energy spectrum corresponding to the enstrophy cascade

$$E(k) \sim \bar{z}k^{-3} \tag{2.7}$$

for the forward cascade range. For the inverse energy cascade we expect the K41 scaling (1.24), since the scaling arguments leading to this scaling nowhere assume a specified direction of the cascade. The arguments above leading to the forward enstrophy cascade and inverse energy cascade were based on the dissipation of enstrophy inhibiting the dissipation of energy at the small scales. Another qualitative argument involving only the nonlinear terms and the inertial flow goes as follows. Consider an experiment where a 2D fluid is excited solely around a wavenumber

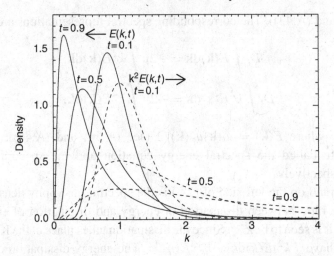

Figure 2.1 The energy initially concentrated around wave number $k_0 = 1$ will spread to neighboring wave numbers through nonlinear wave interactions. The full curves show the energy spectra at different times. These are modeled as log-normal curves with the variance growing linearly in time. The dashed curves are the corresponding enstrophy spectra. The areas under the curves are the energy and enstrophy, respectively. In order to conserve both energy and enstrophy, the energy is cascaded upscale and the enstrophy is cascaded downscale simultaneously.

k_0, suppressing vector notation in the following. The triad interactions will then effectively diffuse the energy at wavenumber k_0 locally under the constraint that the two integrals

$$E = \int E(k)dk, \quad Z = \int k^2 E(k)dk \qquad (2.8)$$

are constant. Since the large k tail will dominate the second integral, this can happen only if the maximum of the curve moves toward smaller wave numbers and upscale energy transfer accompanies downscale enstrophy transfer. This is schematically shown in Figure 2.1.

To understand the mechanism of the dual cascade, consider a vortex of scale R that rotates as a rigid body with rotational speed Ω. Thus the velocity field is $u_i(\mathbf{r}) = \epsilon_{ijl}\Omega_j r_l$ for $r < R$ and $u_i(\mathbf{r}) = 0$ elsewhere, and Ω_i is perpendicular to the plane of the flow. The energy of the vortex is

$$E = \frac{1}{2}\int u^2 d\mathbf{x} = \frac{\pi}{4}\Omega^2 R^4. \qquad (2.9)$$

For the rigid body rotation, using (A.19) we have $\omega_l = \epsilon_{lmn}\partial_m u_n = \epsilon_{lmn}\partial_m(\epsilon_{njk}\Omega_j r_k) = (\delta_{lj}\delta_{mk} - \delta_{lk}\delta_{mj})\Omega_j\partial_m r_k = \Omega_l\partial_k r_k - \Omega_j\partial_j r_l = 2\Omega_l$, and the

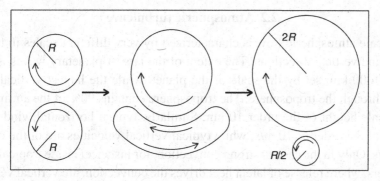

Figure 2.2 The scattering of two vortices of radius R in a 2D flow. The lower vortex is stretched in the flow of the upper vortex. This results in growth to size $2R$ of the upper vortex while a smaller vortex of size $R/2$ is scattered off. If the vortices are considered to perform rigid body rotations, the big upper vortex contains most of the energy while the small lower vortex contains most of the enstrophy. In this way the energy is cascaded to large scales while enstrophy is cascaded to small scales.

enstrophy of the vortex is

$$Z = \int \omega^2 \mathrm{d}x = 4\pi\Omega^2 R^2. \tag{2.10}$$

Consider now a flow of two such vortices of linear size R. Assume that they scatter in a process in which two vortices of linear size, say, $R/2$ and $2R$ emerge. This is schematically shown in Figure 2.2. The picture is similar to what is observed in 2D turbulent flow. These new vortices have rotational speeds Ω_1 and Ω_2, respectively. From energy and enstrophy conservation Ω_1 and Ω_2 are determined:

$$2\frac{\pi}{4}\Omega^2 R^4 = \frac{\pi}{4}\Omega_1^2(R/2)^4 + \frac{\pi}{4}\Omega_2^2(2R)^4,$$

$$8\pi\Omega^2 R^2 = 4\pi\Omega_1^2(R/2)^2 + 4\pi\Omega_2^2(2R)^2, \tag{2.11}$$

from which we get $\Omega_1^2 = 16\Omega^2/5$ and $\Omega_2^2 = \Omega^2/20$. The energy is then redistributed such that

$$E_1 = \frac{4}{5}E \text{ and } E_2 = \frac{1}{5}E, \tag{2.12}$$

while the enstrophy is distributed such that

$$Z_1 = \frac{1}{5}Z \text{ and } Z_2 = \frac{4}{5}Z. \tag{2.13}$$

This means that the energy has moved to larger scales, while the enstrophy has moved to smaller scales.

2.2 Atmospheric turbulence

Large scale atmospheric flow is characterized by very different scales in the horizontal and vertical directions. The extent of the largest planetary waves is of the order 10 000 km, set by the scale of the planet, while the largest vertical waves extend through the troposphere. The troposphere contains 90% of the air mass and has a scale height of the order 10 km. Similarly, typical horizontal wind velocities are of the order of 10 m/s, while typical vertical velocities are of the order of 10^{-2} m/s. Only in the case of strong convection, for instance, inside tropical cumulus towers, where release of latent heat drives the convection, are vertical velocities comparable to horizontal wind velocities. These three orders of magnitude scale differences in the horizontal and vertical directions make atmospheric flow appear two dimensional. The character of atmospheric turbulence strongly influences the possibility of making useful weather predictions. A weather prediction is the initial value problem of calculating the future development of the flow and temperature fields given some initial observations. Obviously, the initial observations are a coarse grained "snapshot" depending on the density of the observational network. For weather prediction to be useful it must specify the conditions down to the scale over which conditions change appreciably. The weather at mid-latitudes is dominated by baroclinic waves of typical scales of 1000 km. These are at mid-latitudes the advected highs and lows typically drifting eastward due to the rotation of the Earth. The weather thus does not change appreciably over scales of, say, a quarter wavelength, or a few hundred kilometers, except across fronts or in places with strong local orographic effects influencing the weather. Changes on scales smaller than this will thus be more or less irrelevant for the weather. Therefore it is not necessary to predict the evolution of small scale events, like the development of a single cumulus cloud, in order to perform a useful weather prediction. Obviously, predicting the development of a single cumulus cloud would limit predictability times to less than hours instead of the several days relevant for weather predictions.

Following Lorenz (1969), a limiting factor for predictability is the back propagation of "errors" in the small scales into the large scales of interest, as described in Section 5.5. An "error" in this context is defined as the difference between the observed fields and the fields which would develop if infinitesimal changes were made in the initial conditions. The origin of these errors can be imperfect initial observations or simply, even with perfect initial knowledge, the effect of the rapid development of chaotic motion in the small scale turbulence. The timescale of this development is many orders of magnitude smaller than the timescale of weather predictions. This short timescale is determined by the Lyapunov exponent which we shall return to in Chapter 5. The time it takes for an error of small length scales to develop into the large length scales is determined by the energy spectrum of the flow. The steeper the slope of the upper tail of the spectrum (large wavenumbers k), the less energy there is in the small scale flow and the less it will influence the large scales. Thus we should expect a 2D turbulent flow with an energy spectrum $E(k) \sim k^{-3}$ to be more predictable in the large scales than a 3D flow with $E(k) \sim k^{-5/3}$. This can also be seen by the heuristic argument presented in (5.20). The timescale for error propagation from scale $(2k)^{-1}$ to, say, scale k^{-1} will be of the order of the eddy turnover time $\tau(k) \sim (k^3 E(k))^{-1/2}$. With $E(k) \sim k^{-\gamma}$ we have $\tau(k) \sim k^{-(3-\gamma)/2}$, and the timescale T_K for error propagation from infinitesimally small scales to the scale K is then

$$T_K \sim \sum_{n=0}^{\infty} \tau(2^n K) = K \sum_{n=0}^{\infty} 2^{n(3-\gamma)/2} = \frac{K}{1 - 2^{(3-\gamma)/2}} \qquad (2.14)$$

for $\gamma < 3$ and $T_K = \infty$ for $\gamma \geq 3$. The dimensionality and thus the spectral slope of the energy spectrum of the atmospheric flow is thus of primary importance to the possibility of performing useful weather predictions.

2.3 The governing equations

The governing equations for atmospheric flow specifying wind velocities, pressure, temperature and density are the Navier–Stokes equation, the continuity equation, the thermodynamic equation and the ideal gas law as described in Section 1.1. The flow is constrained by gravity to the spherical geometry of the planetary surface and is influenced by the rotation of the planet. The rotation results in a centrifugal force and a Coriolis force on the flow, seen from a frame of reference fixed to the surface of the Earth. The centrifugal force is proportional to the distance to, and directed away from, the axis of rotation of the Earth. This makes the Earth ellipsoidal (the geoid), such that the sum of gravity and the centrifugal force is always along the plumb-line. We can thus ignore the centrifugal force due to Earth rotation by simply defining the gravitational acceleration to be along the plumb-line by including the centrifugal force into an effective gravity. The ratio of the centrifugal force to the gravitational force at the equator is of the order $\Omega^2 a/g \approx 0.0035$, a being the radius of the Earth, so the Earth is very close to being spherical. The Coriolis acceleration is $-2\overline{\Omega} \times \mathbf{u}$ per unit mass, where $\overline{\Omega}$ is the rotation vector of the Earth and $\mathbf{u} = (\mathbf{V}, w)$ is the wind velocity. In the last term the velocity is explicitly written in terms of the horizontal velocity vector $\mathbf{V} = (u, v)$ and the vertical velocity w. Defining the usual Cartesian coordinates ($\mathbf{x} = x\mathbf{i} + y\mathbf{j} + z\mathbf{k}$) with x along the latitude circle, y along the longitude circle and z along the vertical, the rotation becomes $\overline{\Omega} = \Omega \cos\phi \mathbf{j} + \Omega \sin\phi \mathbf{k}$, where ϕ is the latitude. The

Coriolis acceleration then becomes $A_C = -(2\Omega \cos\phi\ \mathbf{j} + 2\Omega \sin\phi\ \mathbf{k}) \times (\mathbf{V}, w) = -f\mathbf{k} \times \mathbf{V} - 2\Omega \cos\phi\,(w\mathbf{i} - u\mathbf{k})$. Here we have defined the Coriolis parameter $f = 2\Omega \sin\phi$. Again, the vertical component of the Coriolis force is negligible in comparison to gravity: $2\Omega \cos\phi\ u/g \sim 10^{-4}$ for typical wind velocities. The term $2\Omega \cos\phi\, w$ can be neglected as well since in comparison to the first term we have $w/|\mathbf{V}| \sim 10^{-3}$. The horizontal and vertical momentum equations then read (Holton, 1992)

$$\frac{D\mathbf{V}}{Dt} = \frac{\partial \mathbf{V}}{\partial t} + \mathbf{V} \cdot \nabla \mathbf{V} + w\frac{\partial \mathbf{V}}{\partial z} = -\frac{1}{\rho}\nabla p - f\mathbf{k} \times \mathbf{V}, \qquad (2.15)$$

$$\frac{Dw}{Dt} = \frac{\partial w}{\partial t} + \mathbf{V} \cdot \nabla w + w\frac{\partial w}{\partial z} = -\frac{1}{\rho}\frac{\partial p}{\partial z} - g, \qquad (2.16)$$

where $\nabla = (\partial_x, \partial_y)$ is defined as the horizontal gradient. Comparing this with the Navier–Stokes Equation (1.1), we have neglected the dissipation term, which is many orders of magnitude smaller than any other terms in the equations. The atmospheric flow is characterized by a Reynolds number $\sim 10^5$ or even higher. In general, dissipation due to surface roughness will be the dominant friction, but that too is unimportant for the present discussion. The continuity equation and the equation of state read

$$\nabla \cdot \mathbf{V} + \frac{\partial w}{\partial z} = -\frac{\partial \rho}{\partial t},$$

$$p = \rho R T, \qquad (2.17)$$

where $R = 287$ J/kg K is the gas constant for air. To close the system of equations we finally have the thermodynamic equation for the temperature. This equation is obtained from the first law of thermodynamics, expressing energy conservation. The energy in an air parcel of mass $\delta m = \rho\delta V$ is the sum of the internal energy $\mathcal{E} = (\rho\delta V)c_v T$ and the mechanical energy $(\rho\delta V)u^2/2$. The rate of change of the energy is the sum of the work W done by the forces and the heat $Q = (\rho\delta V)J$ applied to the parcel, where J is the heat per unit mass by radiation, conduction, and latent heat released by condensation. The only forces of meteorological significance in the energy equation are the pressure gradient force and gravity. The Coriolis force is always perpendicular to the direction of the flow, so it cannot do work on the air parcel. The work done is thus $W = -(\nabla \cdot (p\mathbf{u}) + \rho g w)\delta V$, where the last term represents the change in gravitational potential energy. By applying $\mathbf{v}\cdot$ to the NSE, we get the mechanical energy equation $D(\rho u^2/2)/Dt = -\mathbf{u} \cdot \nabla p - \rho g w$, which together with the continuity equation $D_t(\rho\delta V) = D_t\rho + \rho\nabla \cdot \mathbf{u} = 0$ gives the thermodynamic equation

$$c_v \frac{DT}{Dt} = \frac{RT}{\rho}\frac{D\rho}{Dt} + J. \qquad (2.18)$$

This closes the system, which now contains five prognostic equations and one diagnostic equation, the equation of state. A prognostic equation is an equation describing the time evolution of some variables, thus it contains derivatives with respect to time.

2.3.1 Potential temperature

Following an air parcel, initially at the surface $T(0) = T_0$ and $\rho(0) = \rho_0$, the thermodynamic Equation (2.18) can be integrated in time. Assuming no heating ($J = 0$), that is an adiabatic motion, the temperature change is obtained from

$$\int_{T_0}^{T(t)} \frac{D \log T}{Dt} dt = \int_{\rho_0}^{\rho(t)} \frac{R}{c_v} \frac{D \log \rho}{Dt} dt, \qquad (2.19)$$

from which we have $T = T_0(\rho/\rho_0)^{R/c_v}$. Substituting for the density using the equation of state and using $c_p = R + c_v$, this becomes

$$T_0 \equiv \theta = T \left(\frac{p_0}{p} \right)^{R/c_p}. \qquad (2.20)$$

This is defined as the potential temperature when p_0 is the pressure at the surface. The potential temperature signifies the temperature an air parcel of temperature T will have if it is moved adiabatically from the level p to the surface. It is just the Poisson adiabatic state equation, $Tp^{\gamma/(\gamma-1)} = constant$, for the atmosphere, where $\gamma = C_p/C_V$ is the adiabatic exponent.

2.3.2 Hydrostatic balance

The two terms on the right hand side of the vertical momentum Equation (2.16) are both of the order 10 m/s^2, and as such many orders of magnitude larger than both the term on the left hand side and the Coriolis term, which we ignored in comparison to gravity. Imposing an exact balance between these two large terms

$$\frac{\partial p}{\partial z} = -\rho g, \qquad (2.21)$$

the pressure at a point is solely determined by the weight of the air vertically above the point. This is the hydrostatic approximation. It eliminates vertically propagating sound waves and the vertical velocity as a prognostic variable. The vertical velocity then becomes a diagnostic variable, which is set by the requirement of maintaining hydrostatic balance. This set of equations for atmospheric flow is termed the primitive equations.

2.4 Geostrophic wind and the Rossby number

The ratio of the inertial term and the Coriolis term in the momentum equation is characterized by the Rossby number

$$\mathsf{R}o = \frac{(U^2/L)}{fU} = \frac{U}{fL}. \tag{2.22}$$

With a typical wind velocity $U \sim 10$ m/s and horizontal length scale for mid-latitude weather patterns $L \sim 10^6$ m and with the Coriolis parameter $f \sim 10^{-4}$ s^{-1} we have $\mathsf{R}o \sim 0.1$. This means that the term on the left hand side of (2.15) is an order of magnitude smaller than the last term on the right hand side. The only way this can be obtained is through an approximate balance between the two terms on the right hand side. A perfect balance between the two terms would lead to a steady flow given explicitly as

$$\mathbf{V}_g = \frac{1}{f\rho}\mathbf{k} \times \nabla p. \tag{2.23}$$

This is called the geostrophic wind. It is directed along the isobars and approximates to order $\mathcal{O}(\mathsf{R}o)$ the cyclonic wind around highs and lows in the west wind belts.

2.5 Stratification and rotational Froude number

We now introduce the *rotational Froude number* $\mathsf{F}r$, which is defined as

$$\mathsf{F}r = \frac{(fL)^2}{gD}, \tag{2.24}$$

where L and D are typical horizontal and vertical scales, g is the acceleration due to gravity, while f is the Coriolis parameter. The rotational Froude number is a measure of the degree of stratification in a rotating fluid. To see this we take a short detour into the simple case of what is termed the shallow-water case of a rotating fluid (Pedlosky, 1987; Holton, 1992). The fluid is incompressible and has a free surface, but these considerations apply to a stratified atmosphere as well. We now consider linear surface waves: Define the height of the water level as $h(x,y) = D + h'(x,y)$, D being the average depth. Then the horizontal pressure gradient force is, due to change in surface level using hydrostatic balance, given as $\nabla p/\rho = g\nabla h$. Linearizing around $\mathbf{V} = 0$ and $h = D$, which is the trivial solution to the shallow water equation, the horizontal momentum equations become

$$\frac{\partial u}{\partial t} = -g\frac{\partial h'}{\partial x} - fv, \tag{2.25}$$

$$\frac{\partial v}{\partial t} = -g\frac{\partial h'}{\partial y} + fu. \tag{2.26}$$

The first terms on the right hand sides represent the tendency for stratification (flattening the surface) through the influence of gravity, while the second terms represent the influence of rotation of the Earth on the flow. The continuity equation is

$$\frac{\partial h'}{\partial t} = -D\left(\frac{\partial u}{\partial x} + \frac{\partial v}{\partial y}\right). \tag{2.27}$$

Taking the derivative with respect to x of (2.25), taking the derivative with respect to y of (2.26), and adding the two and substituting for u and v using the derivative of (2.27) with respect to time, we obtain the wave equation

$$\frac{\partial^2 h'}{\partial t^2} = -gD\left(\frac{\partial^2 h'}{\partial x^2} + \frac{\partial^2 h'}{\partial y^2}\right) + fD\left(\frac{\partial v}{\partial x} - \frac{\partial u}{\partial y}\right). \tag{2.28}$$

Here we have assumed that the Coriolis parameter f does not change with latitude. This is called the f-plane approximation, valid for $L/a \ll 1$, where a denotes the radius of the Earth. For $f = 0$ the solutions to (2.28) are plane waves with wave velocity $c = \sqrt{gD}$. The scale L_R for appreciable bending of the wave train due to the rotation is given by $L_R f \sim c$, and thus

$$L_R \sim c/f = \sqrt{gD}/f.$$

This is called the Rossby radius of deformation. The (squared) ratio of the horizontal scale of the flow L and the Rossby radius of deformation L_R is recognized from (2.24) as the rotational Froude number $\mathrm{Fr} = L^2/L_R^2$. For $\mathrm{Fr} \ll 1$, flow is dominated by stratification, while for $\mathrm{Fr} \gg 1$, flow is mainly governed by rotation. For the atmosphere we have the scale height $D \sim 10^4$ m and $f \sim 10^{-4}$ s^{-1}, thus $L_R^2 \sim 10^{13}$ m^2. With $L^2 \sim 10^{12}$ m^2 we have $\mathrm{Fr} \sim 10^{-1}$ indicating a strong stratification.

2.6 Quasi-geostrophy

The two dimensionality of the atmosphere follows from a well-controlled approximation termed quasi-geostrophy (Charney, 1971). The relative sizes of the terms in the horizontal momentum Equation (2.15) can be estimated in much the same way as was done for the NSE in (1.6) (Charney & Stern, 1962; Pedlosky, 1987). However, since the fluid now changes density with height, we cannot assume the pressure gradient term to scale in the same way as the inertial term, as it did for the incompressible flow. In order to scale the equations in terms of order unity quantities we have to anticipate the typical sizes of pressure gradients and density in the momentum equation. Writing

$$p = p_s(z) + \tilde{p}(x,y,z,t) \text{ and } \rho = \rho_s(z) + \tilde{\rho}(x,y,z,t), \tag{2.29}$$

with subscript "s" denoting "standard", the first terms on the right hand sides are the horizontal and temporal means. For an atmosphere at rest the hydrostatic balance holds exactly, $dp_s/dz = -g\rho_s$. This relation also holds locally in most meteorologically relevant situations, where vertical velocities are small, and we thus have

$$d\tilde{p}/dz = -g\tilde{\rho}. \tag{2.30}$$

The horizontal pressure gradient is now of the order $\tilde{p}/L$ and using the approximate geostrophic balance ($\mathrm{Ro} \ll 1$) we can scale the pressure using (2.23):

$$\tilde{p} = \rho_s U f_0 L p', \tag{2.31}$$

where p' is dimensionless and of order unity. Here we have introduced the Coriolis parameter at a central latitude ψ_0: $f_0 = 2\Omega \sin \psi_0$. The scaling of the density is obtained from the hydrostatic balance, with D a typical vertical distance over which convection occurs. We have $d\tilde{p}/dz$ of the order $\tilde{p}/D$ and thus $\tilde{\rho} \sim \tilde{p}/(gD)$ which using (2.31) gives

$$\tilde{\rho} = \frac{\rho_s U f_0 L}{gD} p' = \rho_s \, \mathrm{Ro} \, \mathrm{Fr} \rho', \tag{2.32}$$

where ρ' is dimensionless and of order unity. With this we can perform a scale analysis of the horizontal momentum Equation (2.15). Omitting primes for the pressure and density, we arrive at

$$\mathrm{Ro} \frac{D\mathbf{V}}{Dt} = -\frac{1}{1 + \mathrm{Ro} \, \mathrm{Fr} \rho} \nabla p - \frac{\sin \psi}{\sin \psi_0} \mathbf{k} \times \mathbf{V}, \tag{2.33}$$

where all variables are now of order unity. All dimensional quantities are absorbed into the Rossby number, Ro, and the rotational Froude number, Fr. The fraction ($\sin \psi / \sin \psi_0$) accounts for the latitudinal variation of the Coriolis parameter, when we have absorbed the constant f_0 into the Rossby and Froude numbers. Expanding the sine-function around the central latitude yields to first order $\sin \psi = \sin \psi_0 + (L/a)y \cos \psi_0 \equiv (1 + \beta y) \sin \psi_0$. For synoptic scale motion we have $(L/a) \sim 10^{-1}$, which is of the order of the Rossby number so that the latitudinal variation of the Coriolis acceleration cannot be ignored. Proceeding from here, the smallness of the Rossby number is used to expand all the dependent variables in (2.33) in power series of the Rossby number, i.e., $p = p_0 + p_1 \mathrm{Ro} + p_2 \mathrm{Ro}^2 + \ldots$ Inserting this into (2.33) and collecting $\mathcal{O}(1)$ terms we obtain

$$0 = -\nabla p_0 - \mathbf{k} \times \mathbf{V}_0. \tag{2.34}$$

This is simply the geostrophic balance (2.23). To obtain a prognostic equation we must collect the $\mathcal{O}(\mathrm{Ro})$ terms,

$$\frac{D\mathbf{V}_0}{Dt} = -\nabla p_1 + \mathrm{Fr} \, \rho_0 \nabla p_0 - \beta y \mathbf{k} \times \mathbf{V}_0 - \mathbf{k} \times \mathbf{V}_1. \tag{2.35}$$

The continuity and thermodynamic equations can be treated in the same way, and by taking the vertical component of the curl of (2.35), substituting for the vertical velocity, and defining $d/dt \equiv \partial/\partial t + u\partial/\partial x + v\partial/\partial y$, the quasi-geostrophic potential vorticity equation

$$\frac{d}{dt}\left[\zeta_0 + \beta y + \rho_s^{-1}\frac{\partial}{\partial z}\left(\frac{Fr\,\rho_s\theta_0}{\partial \log\theta_s/\partial z}\right)\right] = \frac{\partial}{\partial z}\left(\frac{Fr\,\rho_s J}{\partial \log\theta_s/\partial z}\right) \qquad (2.36)$$

is obtained, where θ is the potential temperature. The derivation of (2.36) is lengthy but straightforward. A detailed account may be found in Pedlosky (1987). The term on the right hand side is a heating term. So, in the case of no heating, the potential vorticity

$$Q = \zeta_0 + \beta y + \rho_s^{-1}\frac{\partial}{\partial z}\left(\frac{Fr\,\rho_s\theta_0}{\partial \log\theta_s/\partial z}\right) \qquad (2.37)$$

is conserved. Thus when the quasi-geostrophic approximation is valid, we can expect the flow to behave like a two-dimensional fluid where both energy and enstrophy are conserved quantities. The enstrophy is here defined as the integral of the quasi-geostrophic potential vorticity.

2.7 Observations of the atmosphere

Kinetic energy spectra of the atmosphere were calculated from observations by Wiin-Nielsen (1967). The spectrum is obtained as the average squared magnitudes of the spherical harmonic expansion coefficients of the observed wind fields (Wiin-Nielsen, 1972; Boer & Shepherd, 1983). It was found that the spectral slope in total wavenumber indeed scales as k^{-3} for synoptic scales as it does in the enstrophy cascading range in 2D turbulence. The resolution obtainable from the observational network at that time limited the scales to larger than total wave numbers around 30, corresponding to spatial scales of the order 30 000 km/30 = 1000 km. Results from an observational campaign measuring wind speeds from commercial aircraft were reported by Nastrom and Gage (1985). With somewhat uneven spatial coverage, the kinetic energy spectra were calculated with resolution down to 1 km. These observations show a crossover from a k^{-3} to a $k^{-5/3}$ spectrum for wavelengths smaller than around 500 km; see Figure 2.3. The origin of this second scaling range is presently not conclusively determined. The horizontal scale, and thus the aspect ratio, is still fairly large such that the two dimensionality of the flow should still hold. This would lead to the conclusion that the range is characterized by an inverse 2D cascade of energy fed at the scale of latent heat release by cumulus convection. Alternatively, the observed spectrum is a result of a forward 3D cascade of energy. In an attempt to observe this range in meteorological fields the observational analysis first done by Wiin-Nielsen (1967) was redone by Straus and Ditlevsen (1999)

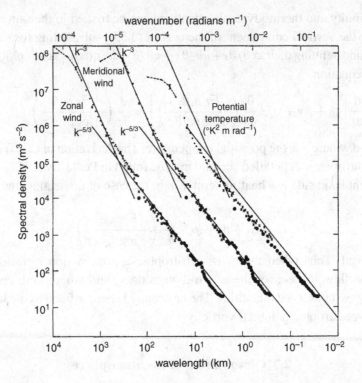

Figure 2.3 Power spectra of wind and potential temperature near the tropopause from GASP (NASA's Global Atmospheric Sampling Program) aircraft measurements. The spectra for meridional wind and potential temperature are shifted by one and two decades to the right in order to separate the data. A crossover from k^{-3} spectra to $k^{-5/3}$ spectra is observed around a few hundred kilometers. (Nastrom & Gage, 1985)

on the higher resolution ECMWF reanalysis dataset (ERA) (Gibson et al., 1997). The reanalysis is a consistent "hind-casting" of 15 years of meteorological data, run through the ECMWF weather forecasting in order to create "best guess" meteorological fields from the observed data. The spectral forecast model was truncated at total wave number $n = 106$, which then corresponds to the smallest possible resolved scale in the fields. The energy spectrum from this analysis is shown in Figure 2.4, where the approximate k^{-3} spectrum is seen. The steepening of the spectrum around $n = 50$ is due to filtering and smoothing by the reanalysis procedure. It is thus not possible to observe any signs of crossover to the $k^{-5/3}$ slope observed by Nastrom and Gage (1985) in the reanalysis data. The spectral fluxes of energy and enstrophy calculated directly from the fields are shown in Figure 2.5.

These show a forward flux of enstrophy (toward smaller scales, larger n) for $n > 30$ and an inverse flux of energy (toward larger scales, smaller n) for $n < 30$.

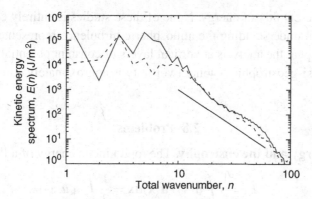

Figure 2.4 The northern hemisphere kinetic energy spectrum in the ECMWF reanalysis data for 15 years (1979–1993). The full curve is the average of the 15 winters, while the dashed curve is for the 15 summers. The kinetic energy in the atmosphere is larger in winter than in summer. (Straus & Ditlevsen, 1999)

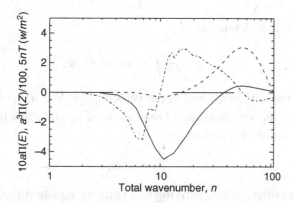

Figure 2.5 The fluxes of energy (full curve) and enstrophy (dashed curve) for the 15 winters. Energy conservation implies energy sources throughout the spectrum (dotted curve). The fluxes and source are rescaled using the radius of the Earth a and the wave number n, in order to plot the three quantities on the same scale. (Straus & Ditlevsen, 1999)

This implies a source of energy at scales around $n = 30$, thus violating an assumption of inertial flow in this range. In the analysis presented here the atmospheric fields are vertically integrated to yield a barotropic (independent of height) flow. The source of energy may well be related to vertical exchange of energy in the atmosphere. This observation does not agree with findings based on analysis of aircraft measurements where third order structure functions were calculated by Lindborg (1999). The sign of these third order structure functions is related to the direction of the fluxes. It was found that the $k^{-5/3}$ range should rather be explained by a 3D forward flux of energy than an inverse 2D flux of energy. None of these studies is entirely conclusive, so a future step in understanding the atmospheric turbulence represented in the data would be to repeat the analysis at vertical layers of constant potential temperature, where the quasi-geostrophic potential vorticity is approximately conserved.

2.8 Problems

2.1 The energy and the enstrophy. The total kinetic energy of a fluid is

$$E = \frac{1}{2}\int \mathbf{u}(\mathbf{x}) \cdot \mathbf{u}(\mathbf{x})d\mathbf{x} = \frac{1}{2}\int u_{\mathbf{k}} u_{-\mathbf{k}} d\mathbf{x}.$$

Show the last identity (Parseval's theorem).
From this you can express the spectral energy density,

$$E = \int E_k dk.$$

The enstrophy is defined as

$$Z = \frac{1}{2}\int \omega(\mathbf{x}) \cdot \omega(\mathbf{x})d\mathbf{x},$$

where the vorticity is the curl of the velocity, $\omega = \nabla \times \mathbf{u}$.
Express the Fourier transform of ω in terms of $u_{\mathbf{k}}$ and show that the spectral enstrophy density is given as

$$Z_k = k^2 E_k.$$

2.2 Vector identities. When deriving the vorticity equation from the NSE we used two vector identities:

$$\nabla \times (\mathbf{A} \times \mathbf{B}) = (\mathbf{B} \cdot \nabla)\mathbf{A} - (\mathbf{A} \cdot \nabla)\mathbf{B} - \mathbf{B}(\nabla \cdot \mathbf{A}) + \mathbf{A}(\nabla \cdot \mathbf{B}),$$

and

$$\frac{1}{2}\nabla \mathbf{u} \cdot \mathbf{u} = (\mathbf{u} \cdot \nabla)\mathbf{u} + \mathbf{u} \times (\nabla \times \mathbf{u}).$$

Prove them using the tensor notation: $\epsilon_{ijk}\partial_j\epsilon_{klm}A_lB_m = \ldots$ Use the relation $\epsilon_{ijk}\epsilon_{klm} = \delta_{il}\delta_{jm} - \delta_{im}\delta_{jl}$.

2.3 Vertical structure of the atmosphere. Using the equation of state (ideal gas) and assuming hydrostatic balance we want to determine the pressure p as a function of height z.
Show that an atmosphere with a constant density must have a finite height. Find this height expressed in terms of surface temperature $T_0 = 300\,\text{K}$, the gas constant $R = 287\,\text{J}/(\text{kg\,K})^{-1}$ and the gravitational acceleration $g = 9.8\,\text{m\,s}^{-2}$.

Determine $p(z)$ for an isothermal atmosphere (constant temperature T). The height $H = RT/g$ is called the scale height of the atmosphere.

Using the expression for the potential temperature (2.20) and hydrostatic balance, derive an expression for the temperature as a function of height in an atmosphere with constant potential temperature. The constant $\Gamma \equiv g/c_p$ is called the (dry) adiabatic lapse rate.

For a dry adiabatic atmosphere, calculate the pressure as a function of height.

3

Shell models

Transfer of energy from large to small scales in turbulent flows is described as a flux of energy from small wave numbers to large wave numbers in the spectral representation of the Navier–Stokes equation (1.17). The problem of resolving the relevant scales in the flow corresponds in the spectral representation to determining the spectral truncation at large wave numbers. The effective number of degrees of freedom in the flow depends on the Reynolds number. The Kolmogorov scale η depends on Reynolds number as $\eta \sim Re^{-3/4}$ (1.11), so the number of waves N necessary to resolve scales larger than η grows with Re as $N \sim \eta^{-3} \sim Re^{9/4}$. This means that even for moderate Reynolds numbers ~ 1000, the effective number of degrees of freedom is of the order of 10^7. A numerical simulation of the Navier–Stokes equation for high Reynolds numbers is therefore impractical without some sort of reduction of the number of degrees of freedom. Such a calculation with a reduced set of waves was first carried out by Lorenz (1972) in the case of the vorticity equation for 2D turbulence.

The idea is to divide the spectral space into concentric spheres, see Figure 3.1. The spheres may be given exponentially growing radii $k_n = \lambda^n$, where $\lambda > 1$ is a constant. The set of wave numbers contained in the nth sphere not contained in the $(n-1)$th sphere is called the nth shell. The nth shell contains of the order of λ^{3n} wave numbers. In shell models only a few wave numbers are retained in each shell. The spectral velocities corresponding to these wave numbers then represent some kind of velocity averaged over the shell. The prefactors of the quadratic terms in (1.17), which are called the interaction coefficients, are chosen such that energy and enstrophy are conserved by the nonlinear interactions. The interaction coefficients couple exactly three waves, since they specify the development of the amplitude of one wave, on the left hand side of (1.17), as a function of two wave amplitudes on the right hand side of (1.17). The procedure of reducing the number of waves has been followed in many subsequent works, and is called reduced wave number models. A rigorous way of reducing the wave numbers in a simulation of

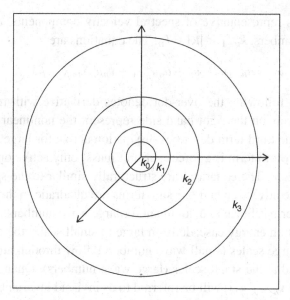

Figure 3.1 The spectral space is divided into spherical shells, each assigned a
wave number which is the radius of the outer sphere.

the Navier–Stokes equation is to impose a symmetry on the forcing and the initial
field. The symmetry must not be violated by the dissipation and advective part
of the equation. As an example, one can imagine a p-fold rotational symmetry,
such that the flow is identical when rotating an angle $2\pi/p$ around some axis. This
will reduce the number of degrees of freedom with a factor of p. The flow can
then retain this symmetry in a simulation even though in general the symmetry
will be spontaneously broken for high Reynolds number flow. Another approach is
the reduced wave vector set approximation (REWA), in which the Navier–Stokes
dynamics is approximated by a set of velocities on wave numbers thinning more
and more for high wave numbers (Grossmann & Lohse, 1994). For a review see
Lesieur (1997).

3.1 The Obukhov shell model

A shell model was first proposed by A. M. Obukhov (1971) from a different perspec-
tive. The model is not derived directly as an approximation of the Navier–Stokes
equation. It is structurally similar, with an energy cascade in accordance with the
Kolmogorov picture of a turbulent cascade of energy. The model is a linear sequence
of coupled first order ordinary differential equations. The equations are nonlinear
and quadratic in a set of velocities, u_n, associated with discrete wave numbers,
$k_n, n = 1, 2, 3, \ldots$ The velocities corresponding to the discrete wave numbers could

be thought of as representative of spectral velocity components, $u_i(\mathbf{k})$, within a shell of wave numbers, $k_{n-1} < |\mathbf{k}| < k_n$. The equations are

$$\dot{u}_n = a_{n-1}u_{n-1}u_n - a_n u_{n+1}^2 - \nu_n u_n \delta_{n>N} + f\delta_{n,1}. \tag{3.1}$$

Here and in the following the over-dot denotes derivative with respect to time. The first two terms on the right hand side represent the nonlinear advection and pressure term, the third term describes dissipation active for large wave numbers $n > N$, and the fourth term is a force (per unit mass) only active on the first wave number component. The equations are structurally similar to the spectral NSE in the sense that the advection and pressure terms are quadratic in the velocities and the dissipation term is linear and dominant for large wave numbers. In order for the model to exhibit an energy cascade from large to small scales the energy must be injected at the large scales (small wave numbers), flow through an inertial range, and be dissipated at the small scales (large wave numbers). Quantities which are conserved when $\nu_n = f = 0$ will be referred to as inviscid invariants. The first two terms conserve the quantity $\sum u_n^2/2$, which will be referred to as the energy E. This can be seen by calculating $\dot{E}$ using (3.1). So for this model the energy is an inviscid invariant. However, the model does not fulfil Liouville's theorem in the inviscid limit. In order to appreciate this we take a short detour into the theory of dynamical systems.

3.2 Liouville's theorem

Liouville's theorem (Goldstein, 1980) is applicable for dynamical systems governed by Hamiltonian equations. Even though the inviscid NSE cannot be put in Hamiltonian form, it is conserving energy and fulfils Liouville's theorem as well. The theorem states that phase space volume is conserved, or equivalently, phase space flow is incompressible. To understand the meaning of this, consider the space of possible initial fields, $u_n(0)$, governed by a first order differential equation like (3.1). In this situation it is useful to think of the subscript n as a usual vector index. This space is called the phase space. Had the governing equation been of second order, such as Newton's law for N particles, the phase space would be spanned by the $2N$ positions and momenta (x_i, p_i) of the particles. In this way, a point in phase space taken as an initial condition determines a unique solution $u_n(t)$, or $(x_i(t), p_i(t))$, the governing equation for all times t, which is a continuous trajectory in phase space. Consider an ensemble of initial fields occupying a volume in phase space. This could be, say, states u_i with energy in the range $[E, E + \Delta E]$. The situation would be relevant if we had only incomplete knowledge of the precise state of a system by having measured only, say, its energy. We then want to know the development of the ensemble of states satisfying our incomplete knowledge of the true state. The phase space flow is described by the tendency vector $\dot{u}_i$, and conservation of phase space volume means that the flow in phase space is incompressible,

$$\partial_i \dot{u}_i = 0, \tag{3.2}$$

where $\partial_i = \partial/\partial u_i$ is the derivative with respect to the phase space coordinates.

Consider the case of N particles in a conservative force field obeying Newton's laws:

$$m_i \ddot{x}_i = \mathbf{F}_i(\mathbf{x}_1,\ldots,\mathbf{x}_N) = -\partial_i V(\mathbf{x}_1,\ldots,\mathbf{x}_N), \tag{3.3}$$

where $V(\mathbf{x}_1,\ldots,\mathbf{x}_N)$ is the potential energy depending on the particle positions, and not their velocities. This is a Hamiltonian system, where the Hamiltonian is the sum of the kinetic and potential energies:

$$H(\mathbf{x}_1,\ldots,\mathbf{x}_N,\mathbf{p}_1,\ldots,\mathbf{p}_N) = \sum_i \frac{p_i^2}{2m_i} + V(x_1,\ldots,x_N), \tag{3.4}$$

where $\mathbf{p}_i = m_i \dot{\mathbf{x}}_i$. The set of $3N$ second order differential equations (3.3) is equivalent to the set of $6N$ first order differential equations:

$$\dot{\mathbf{p}}_i = -\frac{\partial V(\mathbf{x}_1,\ldots,\mathbf{x}_N)}{\partial \mathbf{x}_i} = -\frac{\partial H}{\partial \mathbf{x}_i}, \tag{3.5}$$

$$\dot{\mathbf{x}}_i = \frac{\mathbf{p}_i}{m_i} = \frac{\partial H}{\partial \mathbf{p}_i}. \tag{3.6}$$

These are Hamilton's equations. Liouville's theorem can now easily be proven for this system. In order to do that we rename $\mathbf{x}_1,\ldots,\mathbf{p}_N$ as $\mathbf{y} = (y_1,\ldots,y_{6N})$, and (3.5), (3.6) as

$$\dot{y}_i = G_i(y_1,\ldots,y_{6N}). \tag{3.7}$$

An arbitrary volume $v(t)$ of a region $R(t)$ in phase space is defined as

$$v(t) = \int_{R(t)} dy_1 \ldots dy_{6N}. \tag{3.8}$$

We now want to calculate the volume of the region $R(t+dt)$ at time $t+dt$ occupied by phase space points which originated in $R(t)$ at time t. Any point $\mathbf{y}(t)$ is transformed using (3.7) into a point $\mathbf{y}(t+dt)$ in $R(t+dt)$:

$$y_i(t+dt) = y_i(t) + G_i(\mathbf{y}(t))dt + o(dt^2). \tag{3.9}$$

The volume $v(t+dt)$ of $R(t+dt)$ is now calculated by a simple change of variables from $z_i \equiv y_i(t+dt)$ to $y_i(t)$:

$$v(t+dt) = \int_{R(t+dt)} dz_1 \ldots dz_{6N} = \int_{R(t)} \det(\partial z_i/\partial y_j) dy_1 \ldots dy_{6N}. \tag{3.10}$$

The Jacobian $\det(\partial z_i/\partial y_j)$ is easily evaluated, since $\partial z_i/\partial y_j = \delta_{ij} + \partial_j G_i dt$ to first order in dt only diagonal terms in the determinant contribute:

$$\det(\partial z_i/\partial y_j) = 1 + \partial_i G_i dt, \tag{3.11}$$

and the volume becomes

$$v(t+dt) = v(t) + \int_{R(t)} \partial_j G_j dy_1 \ldots dy_{6N}. \tag{3.12}$$

Returning to Hamilton's equations (3.5) and (3.6), we see that $\partial_i G_i = 0$, since the two terms

$$\partial_1 G_1 = \partial \dot{x}_1/\partial x_1 = \partial^2 H/\partial p_1 \partial x_1$$

and

$$\partial_{3N+1} G_{3N+1} = \partial \dot{p}_1/\partial p_1 = -\partial^2 H/\partial x_1 \partial p_1$$

add to zero (for brevity here x_1 and p_1 are taken to be the first coordinate of the first variable). This shows that $v(t)$ is constant, which proves Liouville's theorem.

3.3 The Gledzer shell model

A set of equations trivially fulfils Liouville's theorem if u_n does not occur in the nonlinear terms of the equation for $\dot{u}_n$. This led E. B. Gledzer (1973) to propose the following set of equations:

$$\dot{u}_n = A_n u_{n+1} u_{n+2} + B_n u_{n-1} u_{n+1} + C_n u_{n-2} u_{n-1} - \nu_n u_n \delta_{n>N_d} + f_n, \tag{3.13}$$

which fulfils Liouville's theorem. As "lower" boundary conditions we have $u_{-1} = u_0 = 0$, and we may or may not specify "upper" boundary conditions similarly as $u_{N+1} = u_{N+2} = 0$.

With this choice of nonlinear interaction terms, the interaction coefficients A_n, B_n, C_n can be chosen such that energy, $E = \sum_n u_n^2/2$, and enstrophy, $Z = \sum_n k_n^2 u_n^2/2$, are inviscid invariants corresponding to 2D turbulence. We return to the inviscid invariants in detail in the next section. The shell model proposed by

Gledzer was investigated numerically by Yamada and Okhitani (1988a), a some 15 years after it was proposed by Gledzer. Their simulations showed that the model exhibits an enstrophy cascade and chaotic dynamics. Interest in shell models grew rapidly after that and many papers investigating shell models have been published since. As there is no direct link between the Navier–Stokes equation and shell models, several shell models have been proposed (Desnyansky & Novikov, 1974). The most well studied model is that proposed by Gledzer and investigated by Yamada and Okhitani. It is now in a complex version referred to as the Gledzer–Okhitani–Yamada or GOY model. We shall later return to a different shell model called the Sabra model[*] by L'vov *et al.* (1998) from the Weissmann Institute in Rehovot, Israel.

3.4 Scale invariance of the shell model

Until this point there are no direct associations between the shell velocities, u_n, and the wave numbers, k_n. By comparing (3.13) and the spectral Navier–Stokes equation (1.17) we can see that the interaction coefficients must have the dimension of wave number, so that we can redefine the interaction coefficients $A_n = k_n a_n$, $B_n = k_n b_n$, $C_n = k_n c_n$, where a_n, b_n, and c_n are dimensionless interaction coefficients. From (1.50) and the spectral NSE (1.17) we find that the Navier–Stokes equation is invariant under the scaling transformation $(t, k, u(k), v) \rightarrow (\lambda^{1-h}t, \lambda^{-1}k, \lambda^{h+D}u(\lambda^{-1}k), \lambda^{1+h}v)$ for any h. The scaling of the spectral velocities with the dimension D arises from the spectral NSE being an integral equation where the volume element scales as $\mathbf{dk} \rightarrow \lambda^{-D}\mathbf{dk}$. The shell model should possess the same type of scale invariance. With $k_m = \lambda^{-1}k_n$ we can apply the transformation $(t, k_n, u_n) \rightarrow (\lambda^{1-h}t, k_m, \lambda^h u_m)$ to (3.13) and obtain, $(v = f = 0)$,

$$\dot{u}_m = k_m a_n u_{m+1} u_{m+2} + k_m b_n u_{m-1} u_{m+1} + k_m c_n u_{m-2} u_{m-1}. \tag{3.14}$$

For this to be identical with the inviscid part of (3.13) the interaction coefficients must be independent of wave number n: $a_n = \tilde{a}, b_n = \tilde{b}$, and $c_n = \tilde{c}$. This should be no surprise since the coefficients are dimensionless and scale invariance is imposed. We can thus rewrite the governing equation to

$$\dot{u}_n = k_n(\tilde{a}u_{n+1}u_{n+2} + \tilde{b}u_{n-1}u_{n+1} + \tilde{c}u_{n-2}u_{n-1}) - vk_n^2 u_n + f_n, \tag{3.15}$$

where we have introduced the dimensionally correct dissipation.

---

[*] The word "sabra" means something like native Israeli Jew in Hebrew, while the word "goy" is a disparaging term from the Talmud meaning non-Jew. Some editors find this an inappropriate joke. However, the name "sabra" model seems to stick.

3.5 The shell spacing and energy conservation

The shell model is constructed such that the energy $E = \sum_n u_n^2/2$ is an inviscid invariant. This can be seen noting that the invariant must be time independent. Using (3.13) with $\nu = f = 0$:

$$\dot{E} = \frac{d}{dt} \sum_n \dot{u}_n^2/2 = \sum_n u_n \dot{u}_n$$

$$= \sum_n k_n(\tilde{a}u_n u_{n+1} u_{n+2} + \tilde{b}u_{n-1}u_n u_{n+1} + \tilde{c}u_{n-2}u_{n-1}u_n)$$

$$= \sum_n (k_n\tilde{a} + k_{n+1}\tilde{b} + k_{n+2}\tilde{c})u_n u_{n+1} u_{n+2} = 0. \tag{3.16}$$

In the last two terms we have changed just the summation label and used the boundary conditions. Since we can have any set of initial velocities we must have

$$k_n\tilde{a} + k_{n+1}\tilde{b} + k_{n+2}\tilde{c} = 0, \tag{3.17}$$

for E to be an inviscid invariant. In order to fully define the model, only the wave numbers remain to be defined. In the spirit of the desire for deriving scaling relations, the most natural choice is to define the wave numbers as

$$k_n = k_0\lambda^n, \tag{3.18}$$

where λ, called the shell spacing, is a real number larger than 1. This means that the spectral space covered by the shells grows exponentially with the shell number n. In this way it is possible to simulate very high Reynolds number "flow" cost efficiently. Inserting this form of the wave numbers into (3.17) gives

$$k_n(\tilde{a} + \tilde{b}\lambda + \tilde{c}\lambda^2) = 0. \tag{3.19}$$

With the further definition $a = \tilde{a}, b = \tilde{b}\lambda$, and $c = \tilde{c}\lambda^2$ we have

$$a + b + c = 0. \tag{3.20}$$

The GOY shell model was originally defined like this. However, we will define the velocities to be complex numbers. The rationale for this is that from a computational point of view it is desirable that the velocity rapidly visits all parts of the attractor, which in the inviscid case is the energy surface, in phase space. We return to the notion of attractors in phase space in Chapter 5. This desired model behavior seems to be achieved when the effective number of degrees of freedom is two per shell rather than one. Note that for the shell velocity at a given shell to change sign it has to vanish in the case of real velocities, while if the velocity is complex it can maintain a constant modulus and change sign for the real part.

A constraint in the spectral form of velocities in a real flow field is $u_i^*(\mathbf{k}) = u_i(-\mathbf{k})$. An analog for this in the case of shell models will be introduced later. The gradient operator ∂_i in the advection and pressure gradient terms transforms into a multiplication operator ιk_i in the spectral representation, so it is natural to put the imaginary unit "ι" as a factor in front of the nonlinear term in (3.15). Furthermore, for complex velocities the energy is defined as $E = \sum_n u_n u_n^*/2$, thus we must have complex conjugations in the governing equation in order to conserve energy in accordance with (3.16). The final form of the GOY model then becomes

$$\dot{u}_n = \iota k_n \left(a u_{n+1} u_{n+2} + \frac{b}{\lambda} u_{n-1} u_{n+1} + \frac{c}{\lambda^2} u_{n-2} u_{n-1} \right)^* - \nu k_n^2 u_n + f_n, \qquad (3.21)$$

where a, b, c are real constants fulfilling (3.20). With a simple change of units we can eliminate one of the three constants, say a, by dividing the equation by a and absorbing it in the unit of time (and in the constants of the linear terms ν and f). It is a tradition in the literature in this case to define $b = -\epsilon$. Using (3.20) we have $c = \epsilon - 1$ and the governing equation can be written in alternative form

$$\dot{u}_n = \iota k_n \left(u_{n+1} u_{n+2} - \frac{\epsilon}{\lambda} u_{n-1} u_{n+1} + \frac{\epsilon - 1}{\lambda^2} u_{n-2} u_{n-1} \right)^* - \nu k_n^2 u_n + f_n. \qquad (3.22)$$

In this form it is transparent that the model is defined by two free parameters ϵ and λ together with the dimensional quantities k_0, ν, f_n, and initial conditions $u_n(0)$.

3.6 Parameter space for the GOY model

With the definition of the GOY model given by (3.21), we can fully characterize the inviscid invariants of the model as a function of the parameters b, c, λ, where we have $a = 1$ by rescaling of time. The GOY model has two conserved integrals (or rather sums), in the case of no forcing and no dissipation ($\nu = f = 0$). We denote the two conserved integrals by

$$E^{(i)} = \sum_{n=1}^{N} E_n^{(i)} = \frac{1}{2} \sum_{n=1}^{N} k_n^{\alpha_i} |u_n|^2 = \frac{1}{2} k_0^{\alpha_i} \sum_{n=1}^{N} \lambda^{n \alpha_i} |u_n|^2, \qquad (3.23)$$

where $i = 1$ or 2. According to the definition (3.22), one of the invariants is energy, the $E = E^{(1)}$ implying that $\alpha_1 = 0$. We will, however, for a moment relax the

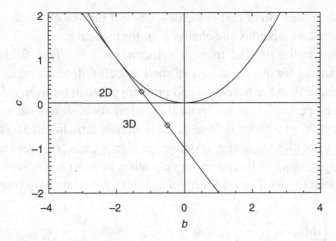

Figure 3.2 The (b, c) parameter space. The line is the set of (b,c) where the one conserved integral is the energy.

constraint $1 + b + c = 0$ given by energy conservation. Setting $\dot{E}^{(i)} = 0$, using $\dot{u}_n$ from (3.21) by exactly the same procedure as in (3.16), we get

$$1 + bz + cz^2 = 0, \tag{3.24}$$

where the roots $z_1 = \lambda^{\alpha_1}$, $z_2 = \lambda^{\alpha_2}$ are called the generators of the conserved integrals. In the case of negative roots z we can use the complex formulation, $\alpha = (\log |z| + \iota \pi)/\log \lambda$. The parameters (b,c) are determined from (3.24) as

$$b = -\frac{z_1 + z_2}{z_1 z_2}, \qquad c = \frac{1}{z_1 z_2}. \tag{3.25}$$

In the (b,c) parameter plane within the (b,c,λ) parameter space the curve $c = b^2/4$ represents models with only one conserved integral, see Figure 3.2. Above the parabola the generators are complex conjugates, and below they are real and different. Any conserved integral represented by a real nonzero generator z defines a line in the (b,c) parameter plane, which is tangent to the parabola at the point $(b,c) = (-2/z, 1/z^2)$.

We now return the focus to the line defined by $1 + b + c = 0$, corresponding to $z_1 = 1$, for which we obtain

$$E^{(1)} = E = \frac{1}{2}\sum_{n=1}^{N} |u_n|^2. \tag{3.26}$$

From here we change the notation to $b = -\epsilon$ and $c = (\epsilon - 1)$, in accordance with (3.22). For further savings in indices we write $z_2 = z$ and $\alpha_2 = \alpha$. The parameter space is now the plane defined by the pair (ϵ, λ).

From (3.25) it is seen that there is another invariant besides the energy invariant $E^{(1)}$. The generator for this other invariant $E^{(2)}$ is

$$z = \frac{1}{\epsilon - 1},$$ (3.27)

and the invariant is

$$E^{(2)} = \frac{1}{2} \sum_{n=1}^{N} \left(\frac{1}{\epsilon - 1}\right)^n |u_n|^2.$$ (3.28)

The model thus has two distinctly different regimes, depending on the sign of z. For $\epsilon < 1$, the invariant $E^{(2)}$ is not positive, and it can be written as

$$E^{(2)} \equiv H = \frac{1}{2} \sum_{n=1}^{N} (-1)^n k_n^{\Re(\alpha)} |u_n|^2,$$ (3.29)

which we will interpret as a generalized helicity.

For $\epsilon > 1$, the second conserved integral is always positive and of the form

$$E^{(2)} \equiv Z = \frac{1}{2} \sum_{n=1}^{N} k_n^{\alpha} |u_n|^2,$$ (3.30)

which we will interpret as a generalized enstrophy. We return to the notion of helicity and enstrophy in the next section.

The (ϵ, λ) phase plane can be reduced using a discrete scaling invariance of the GOY model. By inserting the transformations

$$u'_n = \lambda^{n\beta} u_n,$$
$$t' = \lambda^{-3\beta} t,$$
$$\lambda' = \lambda^{1-\beta},$$
$$\epsilon' = \epsilon \lambda^{2\beta},$$ (3.31)

into the governing Equation (3.22) the equation is left unchanged provided

$$\epsilon' - 1 = (\epsilon - 1)\lambda^{4\beta}.$$ (3.32)

Using (3.31) we obtain the two solutions

$$\lambda^{2\beta} = \begin{cases} 1 \\ 1/(\epsilon - 1) \end{cases},$$ (3.33)

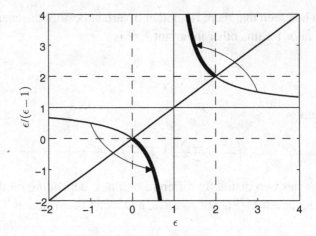

Figure 3.3 The regions $\epsilon < 0$ and $\epsilon > 2$ map onto the region $0 < \epsilon < 2$ by transformation (3.31).

where the first solution is the trivial identity. The second solution transforms the energy into the second invariant and vice versa, that is,

$$(E^{(1)}, E^{(2)})' = (E^{(2)}, E^{(1)}).$$

Inserting the second solution into (3.31) gives the relation $\epsilon' = \epsilon/(\epsilon - 1)$, and as we can see from Figure 3.3, the region $\epsilon < 0$ maps onto the region $0 < \epsilon < 1$ while the region $\epsilon > 2$ maps onto the region $1 < \epsilon < 2$. In the first case the shell spacing becomes $\lambda' = -\iota\lambda\sqrt{1-\epsilon}$ and in the second case $\lambda' = \sqrt{\epsilon - 1}$. The introduction of an imaginary shell spacing in the first case explains why the helicity, which is not positive, can interchange roles with the energy through the transformation (3.31). To conclude this section we can now restrict our interest to the region $0 < \epsilon < 2$ in phase space.

3.7 2D and 3D shell models

The physical dimension of the second inviscid invariant besides energy is determined by the power α of the wave number in (3.23). So from (3.28) we see that the same physical dimension is obtained when α, ϵ, and λ are related through $\lambda^\alpha = |\epsilon - 1|^{-1}$, that is, through the relation

$$\log \lambda = -\frac{1}{\alpha} \log |\epsilon - 1|, \tag{3.34}$$

where the absolute value is used in the denominator in order to cover both the situation where the invariant is of the form (3.29) and of the form (3.30).

As shown in Chapter 6, the helicity

$$H = \int u_i \omega_i \mathrm{d}\mathbf{x} = \int u_i \epsilon_{ijk} \partial_j u_k \mathrm{d}\mathbf{x} \tag{3.35}$$

is an inviscid invariant in 3D flow. The spectral form of the helicity density $h = u_i \omega_i$ is obtained from Fourier transforming the expression (3.35), see (A.10),

$$h(\mathbf{k}) = \iota \epsilon_{ijk} \int (k_j - k_j') u_i(\mathbf{k}) u_k(\mathbf{k} - \mathbf{k}') \mathrm{d}\mathbf{k}'. \tag{3.36}$$

For $\epsilon < 1$, the second invariant is given as $H = \sum_n H_n$ where $H_n = (-1)^n k_n^\alpha |u_n|^2$, so the shell helicity is dimensionally the same as the helicity when $\alpha = 1$ (Kadanoff *et al.*, 1995). The prefactor $\iota \epsilon_{ijk}$ in (3.36) corresponds to the factor $(-1)^n$, reflecting the fact that the helicity in contrast to the energy can be both positive and negative. For $\epsilon < 1$, the shell model is said to be of the 3D turbulence type. We return to interpretation of the helicity in shell models in Section 6.4.

A common choice of parameters for the 3D type GOY model is $(\epsilon, \lambda) = (1/2, 2)$ (Kadanoff *et al.*, 1995), where $\alpha = 1$ and the shell spacing is an octave. The GOY model with these parameter values has been investigated most often and is often referred to as the standard 3D shell model. The "canonical" choice of parameters is nothing but a convenient convention not founded in physical reasoning.

In 2D flow the helicity is trivially conserved. The vorticity is always perpendicular to the plane of the flow, implying that the helicity H is identically zero. The enstrophy is also an inviscid invariant. The spectral density of the enstrophy $Z(k)$ is simply related to the spectral density of energy, $E(k)$, by $Z(k) = k^2 E(k)$, see (A.50). For $\epsilon > 1$, the second invariant in the shell model is $Z = \sum_n Z_n$ with $Z_n = k_n^\alpha |u_n|^2 = k_n^2 E_n$. So for $\alpha = 2$ this corresponds to the enstrophy in 2D flow. The canonical choice in the 2D case is $(\epsilon, \lambda) = (5/4, 2)$ where $\alpha = 2$ (Gledzer, 1973). We see later that 2D shell models are not capable of showing a true inverse energy cascade, which perhaps calls into question the relevance of 2D type shell models as models of 2D turbulence (Aurell *et al.*, 1994; Ditlevsen & Mogensen, 1996).

To summarize, for $\epsilon < 1$ the model is of the 3D type, and for $\epsilon > 1$ the model is of 2D turbulence type. The case $\epsilon = 1$ is the divide between the two turbulence types.

3.8 Other quadratic invariants

Until now we have only considered the possibility of quadratic invariants of the form

$$E^\alpha = \sum_n k_n^\alpha |u_n|^2. \tag{3.37}$$

More generally, there could be invariants of the form

$$E(\mathbf{g}) = \sum_n g_n |u_n|^2, \tag{3.38}$$

where $\mathbf{g} = (g_1, \ldots, g_N)$.

The derivative $d/dt_{n.1}.E(\mathbf{g})$ vanishes if

$$g_{n-1} - \epsilon g_n + (\epsilon - 1)g_{n+1} = 0, \tag{3.39}$$

obtained by use of (3.22) and the same procedure as in (3.16). This is a recursive relation that determines g_n from any set of numbers (g_1, g_2) from which we can eliminate a trivial constant and define $E(\mathbf{g}) = E(g)$ with $g_1 = 1$ and $g_2 = g$. Thus we apparently have an infinity of conservations for the GOY model. However, with this notation the two conserved quantities (3.23) can be represented as $E^{(1)} = E(1)$ and $E^{(2)} = E[1/(\epsilon - 1)]$, and any quadratic conserved quantity can be represented as

$$E(g) = \frac{1 + g(1 - \epsilon)}{2 - \epsilon} E(1) + \frac{(\epsilon - 1)(g - 1)}{2 - \epsilon} E(1/(\epsilon - 1)). \tag{3.40}$$

This shows that there are indeed only two linearly independent quadratic invariants of the shell model.

3.9 Triad interactions and nonlinear fluxes

The shell models are constructed to have detailed energy balance within triads of interacting waves similar to the spectral Navier–Stokes equation. To study this we define the so-called correlators

$$\Delta_n = k_{n-1} \Im(u_{n-1} u_n u_{n+1}). \tag{3.41}$$

Taking the derivative of the energy $E = \sum_n u_n u_n^* / 2$ with respect to time using (3.22), and considering only those terms that involve the shells numbers $n - 1, n, n + 1$, gives

$$\dot{E}_{n-1} = \Delta_n,$$
$$\dot{E}_n = -\epsilon \Delta_n,$$
$$\dot{E}_{n+1} = (\epsilon - 1)\Delta_n. \tag{3.42}$$

It follows from these equations that we have $\dot{E}_{n-1} + \dot{E}_{n-1} + \dot{E}_{n-1} = 0$. In the same way all other terms sum to zero in sets of three involving three consecutive shells.

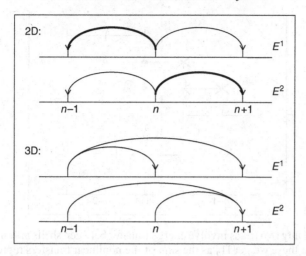

Figure 3.4 The shell triad interactions. Arrows indicate transfer of energy, E, and enstrophy, Z, for the 2D case, $\epsilon = 5/4$; and energy, E, and helicity, H, for the 3D, $\epsilon = 1/2$, case. The thickness of the arrows indicates the strength of the transfer.

Similarly, for the helicity/enstrophy $E^{(2)} = \sum_n k_n^\alpha E_n$:

$$\dot{E}_{n-1}^{(2)} = k_{n-1}^\alpha \Delta_n,$$
$$\dot{E}_n^{(2)} = -[\epsilon/(\epsilon - 1)]k_{n-1}^\alpha \Delta_n,$$
$$\dot{E}_{n+1}^{(2)} = [1/(\epsilon - 1)]k_{n-1}^\alpha \Delta_n, \tag{3.43}$$

and it follows from these equations that we have $\dot{E}_{n-1}^{(2)} + \dot{E}_{n-1}^{(2)} + \dot{E}_{n-1}^{(2)} = 0$ for the terms involving the three shell numbers $n - 1, n, n + 1$.

Figure 3.4 shows the two cases $\epsilon < 1$ for 3D like models and $\epsilon > 1$ for 2D like models. The thickness of the arrows symbolizes the relative sizes of the exchanges in the cases of $\epsilon = 1/2$ and $\epsilon = 5/4$. The actual transfer of energy and helicity/enstrophy depends on the sign of Δ_n; if $\Delta_n < 0$ the arrows in the figure are turned the opposite way. If the transfer is as indicated in the average, there will be a transfer of energy and helicity from large to small scales (small wave numbers to large wave numbers) in the 3D case. In the 2D case we expect that there will be a transfer of enstrophy from large to small scales together with a transfer of energy in the opposite direction, that is, from small to large scales.

The nonlinear flux of energy through a shell n is defined as

$$\Pi_n = \frac{\mathrm{d}}{\mathrm{d}t}\bigg|_{\mathrm{n.l.}} \sum_{m \le n} E_m, \tag{3.44}$$

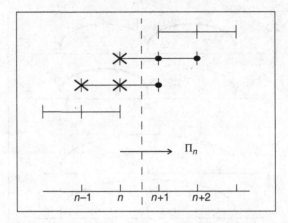

Figure 3.5 Only two triads involve energy transfer between shells $m < n$ and shells $m \geq n$. From these we get Π_n as the sum of the nonlinear transfers represented by the three crosses, which is equal and opposite in sign to the transfers represented by the three dots.

where the derivative $\mathrm{d}/\mathrm{d}t_{\mathrm{n.l.}}$ denotes the rate of change due to the nonlinear interactions. From (3.42) and Figure 3.5 it is easy to obtain the expression

$$\Pi_n = \Delta_{n+1} - (\epsilon - 1)\Delta_n \tag{3.45}$$

for the energy flux by observing that only triads containing both shell numbers larger than and shell numbers smaller than n contribute to the flux. Equivalently, (3.43) gives the expression

$$\Pi_n^{(2)} = \frac{\mathrm{d}}{\mathrm{d}t}_{\mathrm{n.l.}} \sum_{m \leq n} E_m^{(2)} = \left(\frac{1}{\epsilon - 1}\right)^n (\Delta_{n+1} - \Delta_n) \tag{3.46}$$

for the flux of helicity/enstrophy. This expression becomes singular for $\epsilon = 1$, so this case must be treated separately.

3.10 The special case $\epsilon = 1$

The value $\epsilon = 1$ separates the 2D and 3D cases. In the limit $\epsilon \nearrow 1$, the helicity will be more and more concentrated in the large wave number shells. This is a simple consequence of the relation $\lambda^\alpha = 1/(\epsilon - 1)$. The ratio of helicities at consecutive shells is

$$\frac{H_n}{H_{n-1}} = \frac{(-1)^n k_n \alpha E_n}{(-1)^{n-1} k_{n-1} E_{n-1}} = -\lambda^\alpha \frac{E_n}{E_{n-1}}. \tag{3.47}$$

We may assume that E_n/E_{n-1} remains finite and different from zero in the limit $\epsilon \nearrow 1$. In the case $\epsilon = 1$, all the helicity will be concentrated in the outmost shell N, since $\lambda^\alpha \to \infty$ as $\epsilon \nearrow 1$. We can easily convince ourselves by looking at the governing equation

$$\dot{u}_n = \iota k_n u_{n+1}(u_{n+2} - u_{n-1}/\lambda) - \nu k_n^2 u_n + f \delta_{nn_0}, \tag{3.48}$$

corresponding to $\epsilon = 1$, where $\delta_{nn_0} = 1$ for $n = n_0$ and zero otherwise. For the outmost shell we have the equation $\dot{u}_N = -\nu k_N^2 u_N$, which means that the velocity u_N is constant in time in the inviscid case $\nu = 0$. The same argument holds in the limit $\epsilon \searrow 1$, and we see that the helicity and the enstrophy conservations coincide in the conservation of the velocity (squared) at the outmost shell. The behavior of the model for $\epsilon = 1$ is trivial. From (3.48) we see that u_N decreases exponentially to zero due to the viscosity because there are no triads involving u_N from which the Nth shell can obtain energy. Thus when $u_N = 0$ the second last shell must loose all its energy, $u_{N-1} = 0$, and so on. The model will thus run into a fixed point where all velocities vanish except for the shell where energy is pumped in through the force. At this shell a balance between input and dissipation of energy is obtained for $u_{n_0} = f/\nu k_{n_0}^2$. So the fixed point is $\mathbf{u} = (0,\ldots,0,f/\nu k_{n_0}^2,0,\ldots,0)$. The fixed point is stable for all values of f and $\nu > 0$. This can be seen by a linear stability analysis. For $n \geq n_0$ the dissipation term is the only term which is first order in the perturbations, thus the perturbation dies out. For $n < n_0$ we need to consider only the shells $n_0 - 2, n_0 - 1$, from which we get the linearized equations

$$\dot{u}_1 = -\nu_1 u_1 + \iota a u_2,$$
$$\dot{u}_2 = -\iota a u_1 - \nu_2 u_2, \tag{3.49}$$

where we have used the short notation

$$u_i = u_{n_0-i}, \quad a = -k_{n_0-1}f/\nu k_{n_0}^2, \quad \nu_i = \nu k_{n_0-i}, \quad i = 1,2.$$

Multiplying the first equation by u_1, the second equation by u_2 and adding we get

$$\frac{d}{dt}(u_1^2 + u_2^2) = -\nu_1 u_1^2 - \nu_2 u_2^2, \tag{3.50}$$

showing that the fixed point is stable. Note that the term in the bracket on the left side is not the energy $E = u_1 u_1^* + u_2 u_2^*$ in the two shells.

3.11 The Sabra shell model

The nonlinear part of the spectral Navier–Stokes Equation (1.18) will not only conserve energy and helicity (in the 3D case) globally but in each triad involving wave numbers $\mathbf{k}, \mathbf{k}', \mathbf{k}''$, where $\mathbf{k} + \mathbf{k}' + \mathbf{k}'' = 0$. This is the detailed balance, see Appendix A.7.

The Sabra shell model is as before defined by a set of exponentially spaced one-dimensional wave numbers $k_n = k_0 \lambda^n$, for which we have u_n as the complex shell velocity for shell $n \geq 1$ (L'vov *et al.*, 1998; Ditlevsen, 2000). The form of the governing equation for this model is motivated by the demand that the momenta involved in the triad interactions must add up to zero as in the NSE. This, together with the usual construction of local interactions in k-space, inviscid conservation of energy, and fulfilment of Liouville's theorem, gives the equation of motion for the shell velocities

$$\dot{u}_n = \iota [k_n u_{n+1}^* u_{n+2} - \epsilon k_{n-1} u_{n-1}^* u_{n+1} + (1 - \epsilon) k_{n-2} u_{n-2} u_{n-1}]$$
$$- \nu k_n^2 u_n + f_n, \tag{3.51}$$

where ν is the viscosity and f_n is the external force. The force would, as in the GOY model, typically be taken to be active only for some small wave numbers, e.g., $f_n = f \delta_{n,4}$. Boundary conditions can be specified in the usual way by the assignment $u_{-1} = u_0 \, (= u_{N+1} = u_{N+2}) = 0$.

The requirement of closing the triads is fulfilled if the wave numbers k_n are defined as a Fibonacci sequence, $k_n = k_{n-1} + k_{n-2}$. The choice of a Fibonacci sequence for the momenta leads to a model with the shell spacing uniquely being the golden ratio $g = (\sqrt{5} + 1)/2$, since for any choice of $k_1, k_2 \, (k_1 \leq k_2)$ we have $k_n / k_{n-1} \to g$ as $n \to \infty$. So this corresponds to the usual definitions of the shell wave numbers with the golden ratio as the shell spacing for $(k_1, k_2) = (1, g)$. The golden ratio g and $-1/g$ are the roots of the equation $\lambda^2 = \lambda + 1$, so it plays a key role in the symmetries of the shell models.

With this formulation the shell spacing is not a free parameter of the shell model. However, using the definition by L'vov *et al.* (1998) of $k_n = g^n$ being a *quasi-momentum*, we shall keep the shell spacing λ as a free parameter, $k_n = \lambda^n$. The equation of motion can thus be written in the same form as (3.22) for the GOY model:

$$\dot{u}_n = \iota k_n \left(u_{n+1}^* u_{n+2} - \frac{\epsilon}{\lambda} u_{n-1}^* u_{n+1} - \frac{\epsilon - 1}{\lambda^2} u_{n-2} u_{n-1} \right)$$
$$- \nu k_n^2 u_n + f_n. \tag{3.52}$$

With this formulation it is readily seen that the form of the fluxes Π_n becomes similar to (3.45) for the GOY model, except that the correlator becomes

$$\Delta_n = k_{n-1} \Im m (u_{n-1}^* u_n^* u_{n+1}), \tag{3.53}$$

instead of the definition (3.41) for the GOY model.

If we interpret the momentum, k_n, as representative of the modulus of the wave vector, $\mathbf{k}_n$, in 2D or 3D, the triangle inequality implies $k_n + k_{n+1} \geq k_{n+2}$, so the Fibonacci sequence corresponds in this sense to moduli of three parallel wave vectors. Note that for a shell spacing $\lambda > g$ (as the usual choice, $\lambda = 2$) the triangle inequality is violated. This means that we cannot interpret the usual shell model interactions as representative interactions between waves within three consecutive shells, since no such triplets of wave numbers constitute triangles. In this case we must refer to the quasi-momenta.

In order to make sense of the notion of closing the triads, we can define negative momenta, $k_{-n} \equiv -k_n$, and assign the velocity, $u_{-n} = u_n^*$, to these momenta. (The model still has only $2N$ degrees of freedom, represented by the N complex velocities.) It can be seen that (3.51) is also fulfilled for the negative momenta, requiring that the prefactor be "ι". With this notation, we can rewrite (3.51) as

$$(\mathrm{d}/\mathrm{d}t + vk_n^2)u_n = \iota k_n \sum_{k_l < k_m} I(l,m;n)u_l u_m + f_n, \tag{3.54}$$

where the sum is over positive and negative momenta, and all the dimensionless interaction coefficients have the form

$$I(l,m;n) = \delta_{n+1,l}\delta_{n+2,m} - \frac{\epsilon}{\lambda}\delta_{n-1,l}\delta_{n+1,m} + \frac{1-\epsilon}{\lambda}\delta_{n-2,l}\delta_{n-1,m}.$$

From this formulation it becomes evident why the complex conjugations in the Sabra model are exactly as in (3.51). The first term on the right hand side of (3.54) corresponds to the triad where $k_n = -k_{n+1} + k_{n+2}$, and therefore u_{n+1}^* corresponding to wave number $-k_{n+1}$ is in (3.51). The second term corresponds to the triad where $k_n = -k_{n-1} + k_{n+1}$ and therefore u_{n-1}^* appears, while the last term corresponds to the triad $k_n = k_{n-2} + k_{n-1}$, where all wave numbers appearing are positive. Thus this coupling of three wave numbers is similar to the case for the discrete spectral NSE (1.18).

3.12 Problems

3.1 **The golden ratio.** A line segment $[0a]$ is split into two parts $[0b]$ and $[ba]$, such that the ratio of the largest to the smallest segment equals the ratio of the whole to the largest segment. Show that a/b is the golden ratio g.

3.2 **The Fibonacci sequence.** Find for any Fibonacci sequence, defined by $a_1 = 1$, $a_2 = b$, and $a_{n+1} = a_{n-1} + a_n$, the limit for $n \to \infty$ of the ratio a_{n+1}/a_n. Can b be chosen such that $a_n \to 0$ for $n \to \infty$?

3.3 **Recursive map.** Consider the recursive relation (3.39),

$$g_{n-1} - \varepsilon g_n + (\varepsilon - 1)g_{n+1} = 0.$$

By defining $a_n = g_n/g_{n-1}$ derive a map for a_n. Find the fixed points for the map. Determine the stability of the fixed points and interpret these in terms of the conserved quantities of the shell model.

3.4 Show that any quadratic conserved quantity can be expressed by (3.40).

3.5 Show that by defining $u_n = iv_n$, with v_n real velocities, transforms the Sabra model to the real version of the GOY model.

3.6 **A shell model for passive advection.** Consider a scalar field $\theta(\mathbf{x}, t)$ passively carried by a turbulent non-divergent velocity field $\mathbf{u}(\mathbf{x}, t)$. The velocity field evolves according to the NSE,

$$\partial_t u_i + u_j \partial_j u_i = -\partial_i p + \nu \partial_{jj} u_i + f_u.$$

The time evolution of the passive field is governed by the equation:

$$\partial_t \theta + u_i \partial_i \theta = \kappa \partial_{ii} \theta + f_\theta, \tag{3.55}$$

where κ is the diffusion coefficient and f_θ is a source term.

Let u_n be shell velocities for a (GOY or Sabra) shell model for the velocity. Construct from Equation (3.55) a shell model θ_n with nearest and next-nearest neighbor interactions for the scalar field. The model should fulfil the condition that $T = \sum_n |\theta_n|^2$ is constant for $\kappa = f_\theta = 0$.

Determine from dimensional arguments the scaling properties of this model. Compare with numerical simulations.

If you want to learn more about this model, consult (Jensen *et al.*, 1992).

4

Scaling and symmetries

Regardless of the realism of shell models, they can be subjected to exactly the same kind of analysis as was previously used for the dynamics of the NSE. The shell models have the advantage over the NSE that their simplicity makes the analysis much more transparent. With the two inviscid invariants, energy and helicity/enstrophy, in the model, we can now repeat the Kolmogorov dimensional analysis.

4.1 The nonlinear fluxes

In 3D turbulence there is a forward cascade of energy, while in 2D turbulence there is a forward cascade of enstrophy and an inverse cascade of energy. So we do not know in advance the behavior of the two inviscid invariants. We therefore assume that the Kolmogorov hypothesis can be applied to either of the invariants denoted $E^{(i)}$, $i = 1$ or 2. In analogy to turbulence, we define the inertial sub-range as the range of shells where the force and the dissipation are negligible in comparison with the nonlinear interactions between shells. In this range, the spectrum of $E^{(i)}$, by the Kolmogorov hypothesis, depends only on k and $\bar{\varepsilon}_i$, where $\bar{\varepsilon}_i$ is the time averaged dissipation of $E^{(i)}$. From dimensional analysis we have $[ku] = \mathrm{s}^{-1}$, $[\bar{\varepsilon}_i] = [E^{(i)}]\mathrm{s}^{-1}$, $[E^{(i)}] = [k^{\alpha_i} u^2] = [k]^{\alpha_i - 2}\mathrm{s}^{-2}$, and we get

$$E^{(i)} \sim \bar{\varepsilon}_i^{2/3} k^{[\Re e(\alpha_i) - 2]/3}. \tag{4.1}$$

For the velocity u we then get the "Kolmogorov-scaling"

$$|u| \sim \bar{\varepsilon}_i^{1/3} k^{-[\Re e(\alpha_i) + 1]/3}. \tag{4.2}$$

To make this analysis general we can express the nonlinear flux of the conserved quantities (3.44) and (3.46) using the generators $z_1 = 1$ and $z_2 = 1/(\epsilon - 1)$

65

defined from (3.24). The nonlinear flux through shell number n is then expressed directly as

$$\Pi_n^{(1)} = z_1^n(-\Delta_n/z_2 + \Delta_{n+1}),$$
$$\Pi_n^{(2)} = z_2^n(-\Delta_n/z_1 + \Delta_{n+1}). \tag{4.3}$$

In the inertial range the cascade is constant, $\Pi_n^{(i)} = \Pi_{n+1}^{(i)}$, so we get

$$z_1 z_2 \Delta_{n+2} - (z_1 + z_2)\Delta_{n+1} + \Delta_n = 0. \tag{4.4}$$

This equation has the scaling solution $u_n \sim k_n^\gamma$ and thus $\Delta_n \sim k_n^{3\gamma+1}$. Inserting this and using $k_n = \lambda^n$ we get the equation

$$z_1 z_2 (\lambda^{3\gamma+1})^2 - (z_1 + z_2)\lambda^{3\gamma+1} + 1 = 0, \tag{4.5}$$

with the two solutions $\gamma_i = -(\alpha_i + 1)/3$, $i = 1$ and 2, which for the cascade of $E^{(1)}$ is

$$\Pi_n^{(1)} = \Pi^{(1)} \sim \begin{cases} 1 - z_2/z_1 & \text{Kolmogorov for } E^{(1)} \\ 0 & \text{fluxless for } E^{(1)}, \end{cases} \tag{4.6}$$

and correspondingly for $E^{(2)}$

$$\Pi_n^{(2)} = \Pi^{(2)} \sim \begin{cases} 0 & \text{fluxless for } E^{(2)} \\ 1 - z_1/z_2 & \text{Kolmogorov for } E^{(2)}. \end{cases} \tag{4.7}$$

These are the two scaling fixed points for the model. The Kolmogorov fixed point for the first conserved integral corresponds to the fluxless fixed point for the other conserved integral and vice versa. This is of course just a reflection of the fact that (4.5) is symmetric in the indices 1 and 2. In the case of no force and dissipation it is trivial that these points are fixed points in phase space, because $\Pi_n = \Pi_{n+1}$ implies that $\dot{E}_{n+1} = 0$ and thus $\dot{u}_{n+1} = 0$. As it should be, the Kolmogorov fixed point

$$u \sim k^{-(\alpha+1)/3} \tag{4.8}$$

obtained from this analysis is in agreement with the dimensional analysis (4.1).

The scaling fixed points can be obtained directly from the dynamical equation as well. By inserting $u_n \sim k_n^{-\gamma} = \lambda^{-n\gamma}$ into (3.21) with $a = 1$ and $v = f_n = 0$ we get

$$\lambda^{n(1-2\gamma)-3\gamma}[1 + b\lambda^{3\gamma-1} + c(\lambda^{3\gamma-1})^2] = 0, \tag{4.9}$$

and the generators reemerge: $z_i = \lambda^{\alpha_i} = \lambda^{3\gamma_i-1}$. This gives the Kolmogorov fixed points for the two conserved integrals, $\gamma_i = (\alpha_i + 1)/3$. In the case of the Sabra model the fixed point is imaginary and given by $u \sim \iota k^{-(\alpha+1)/3}$, which changes the sign in the last term in (3.52).

4.2 3D GOY and Sabra models

The governing Equations (3.22) for the GOY model and (3.52) for the Sabra model can be solved numerically by any time stepping routine capable of handling stiff systems. A stiff system of ODEs is a system in which there are variations over many different time scales. However, due to the relatively few degrees of freedom of the shell models they can be integrated using a simple fourth order Runge–Kutta integration scheme or the like. Such integrations are shown for the GOY model in Figure 4.1 and for the Sabra model in Figure 4.2. Both models are integrated for $\epsilon = 1/2$ corresponding to 3D turbulence. As we expect from dimensional analysis of the Sabra model, a log-log plot (Figure 4.3) of the spectrum $|u|$ versus k shows that there is a "Kolmogorov-scaling" corresponding to (4.2) for the energy $E^{(1)} = E\,(\alpha_1 = 0)$.

In general the fixed points will not be stable and an equation like (4.4) would hold only in the mean. This then implies that correlation functions $\langle \Delta \rangle$ should be substituted for the Δ variables. Instead of obtaining the fixed points by requiring a steady flux in the inertial range, we can get them directly from the governing equations (3.22) and (3.52) with $\nu = f = 0$.

By rewriting the equation in terms of modulus and a phase $u_n = r_n \exp(\iota \theta_n)$, we have for the GOY model

$$\dot{r}_n + \iota r_n \dot{\theta}_n = k_n \left(r_{n+1} r_{n+2} e^{\iota\,(\pi/2 - \theta_n - \theta_{n+1} - \theta_{n+2})} \right.$$

$$- \frac{\epsilon}{\lambda} r_{n-1} r_{n+1} e^{\iota\,(\pi/2 - \theta_{n-1} - \theta_n - \theta_{n+1})}$$

$$\left. + \frac{\epsilon - 1}{\lambda^2} r_{n-2} r_{n-1} e^{\iota\,(\pi/2 - \theta_{n-2} - \theta_{n-1} - \theta_n)} \right)$$

$$- \nu k_n^2 r_n + f_n e^{-\iota\theta_n}, \qquad (4.10)$$

where we have divided both sides by $\exp(\iota\,\theta_n)$. If we choose $\tilde{f}_n = f_n e^{-\iota\theta_n}$ to be real, the right hand side of the equation is real provided that $\theta_n + \theta_{n+1} + \theta_{n+2} = \pi/2$ for all n. This implies that, if r_n is a fixed point of

$$\dot{r}_n = k_n \left(r_{n+1} r_{n+2} - \frac{\epsilon}{\lambda} r_{n-1} r_{n+1} + \frac{\epsilon - 1}{\lambda^2} r_{n-2} r_{n-1} \right) - \nu k_n^2 r_n + \tilde{f}_n, \qquad (4.11)$$

then $u_n = r_n \exp(\iota \theta_n)$ with

$$\theta_n = \begin{cases} \phi & \text{for } n = 0 \bmod 3 \\ \psi & \text{for } n = 1 \bmod 3, \\ \pi/2 - \phi - \psi & \text{for } n = 2 \bmod 3 \end{cases} \qquad (4.12)$$

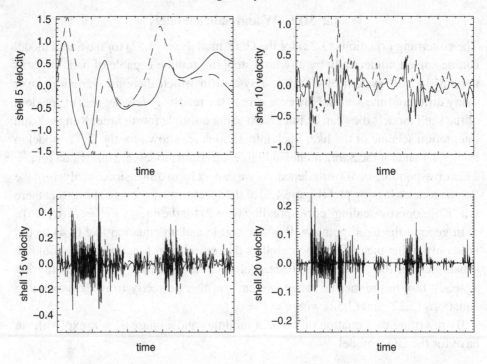

Figure 4.1 A numerical realization of the GOY model. The full lines are the real part of the shell velocities, while the broken lines are the imaginary part of the shell velocities. Shell numbers 5, 10, 15 are in the inertial range, while shell number 20 is in the dissipative range. The last shell shows intermittent bursts of energy arriving at the dissipative scale.

is a fixed point of (3.22). The corresponding equation in terms of modulus and phase for the Sabra model reads

$$
\begin{aligned}
\dot{r}_n + \iota\, r_n \dot{\theta}_n = k_n \Big(& r_{n+1} r_{n+2} e^{\iota(\pi/2 - \theta_n - \theta_{n+1} + \theta_{n+2})} \\
& - \frac{\epsilon}{\lambda} r_{n-1} r_{n+1} e^{\iota(\pi/2 - \theta_{n-1} - \theta_n + \theta_{n+1})} \\
& + \frac{\epsilon - 1}{\lambda^2} r_{n-2} r_{n-1} e^{\iota(-\pi/2 + \theta_{n-2} + \theta_{n-1} - \theta_n)} \Big) \\
& - \nu k_n^2 r_n + f_n e^{-\iota \theta_n},
\end{aligned}
\tag{4.13}
$$

which results in the same real Equation (4.11) provided that

$$
\theta_{n+1} = \theta_{n-1} + \theta_{n+1} - \pi/2.
\tag{4.14}
$$

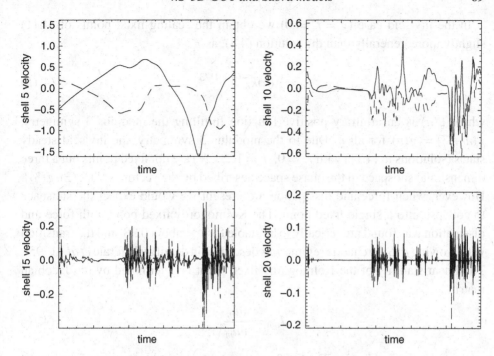

Figure 4.2 Same as Figure 4.1, but for the Sabra model.

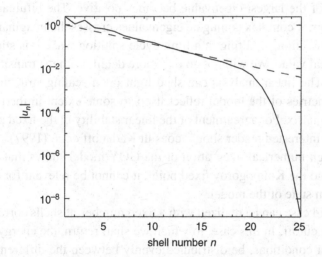

Figure 4.3 The spectrum $\log|u|$ as a function of $n \sim \log k$ for the Sabra model. This is the Kolmogorov scaling expected from dimensional analysis. The linear regime is the inertial range, where the slope is the scaling exponent, which is close to the K41 value 1/3 shown by the dashed line.

In the inviscid case ($\nu = f_n = 0$) we obtain the scaling fixed points of (4.11) slightly more generally than the solution (4.8), as

$$r_n = g(n)k_n^{-(\alpha_i+1)/3}, \tag{4.15}$$

where $g(n)$ is an arbitrary positive function fulfilling the modulus 3 symmetry, $g(n+3) = g(n)$ for all n. Due to the modulus 3 symmetry, the inviscid steady states (solutions to (4.11) with $\dot{r}_n = 0$, $n = 1,\ldots,N$) are not fixed points but a three dimensional subspace in the phase space described by the vector $[g(1), g(2), g(3)]$. However, when force and dissipation are present we would expect the subspace to collapse into a single fixed point. The Kolmogorov fixed point with force and dissipation was found numerically to be stable for $\epsilon <$ about 0.38, and the transition with growing ϵ to the chaotic regime was described in detail by Biferale *et al.* (1995). The linear stability of the Kolmogorov fixed point is determined by the Jacobian matrix

$$J_{nm} = \frac{\partial \dot{u}_n}{\partial u_m}, \tag{4.16}$$

evaluated at the fixed point. The fixed point is stable when the real part of the largest eigenvalue is negative. As a function of ϵ, a bifurcation point is obtained where the real part of the largest eigenvalue becomes positive. The bifurcation point is such that a pair of complex conjugate eigenvalues crosses the imaginary axis. This is a Hopf bifurcation, resulting in a limit cycle solution when ϵ is slightly bigger than the critical value. We will not go into more details with the transition to chaos in this case. The linear analysis can shed light upon scaling structure, since the scaling symmetries of the model reflect itself to some extent in the properties of the Jacobian matrix. For a treatment of the linear stability of the fixed point and the spectrum the interested reader should consult Kadanoff *et al.* (1997).

Even though numerical integration of the GOY model shows that the solution evolves around the Kolmogorov fixed point, it cannot be relevant for determining the time-mean state of the model.

The inviscid case can be realized with a fixed number of shells corresponding to an ultraviolet cutoff. In this case, to which we shall return, the energy will, under rather general conditions, be distributed evenly between the different degrees of freedom of the system. This is called equipartition of the energy . It implies that $\langle E_n \rangle = \langle |u_n|^2 \rangle / 2 = kT$ and the scaling becomes $|u_n| \sim$ const. This state is far away from the Kolmogorov scaling $|u_n| \sim k_n^{-1/3}$, and it is a shell model manifestation of inviscid flow (governed by the Euler equation) not being the inviscid limit of the Navier–Stokes equation. The Euler equation is identical to the NSE without a viscous term.

The shell models also have a fixed point for low Reynolds numbers more relevant to fluid flow. This corresponds to the transition from a steady laminar flow to turbulence.

The stability of the last fixed point, the trivial fixed point $\mathbf{u} = 0$, cannot be determined by a linear stability analysis in the inviscid and unforced case. That is because the Jacobian vanishes for $\mathbf{u} = 0$ so that the second order expansion becomes the governing equation itself. From energy conservation, the stability must be determined by the linear terms. If the force is absent, the trivial fixed point will always be stable, corresponding to the end state of decaying turbulence.

4.3 Phase symmetries

The statistics of the shell models are expressed through the means of moments of the shell velocities and correlations between velocities at different shells. These correspond to the structure functions for real fluids. The most general structure functions can be defined as

$$S_{q_1,q_2,\ldots;p_1,p_2,\ldots}(n_1,n_2,\ldots;m_1,m_2,\ldots) = \langle u_{n_1}^{q_1} u_{n_2}^{q_2} \ldots u_{m_1}^{*p_1} u_{m_2}^{*p_2} \ldots \rangle. \tag{4.17}$$

Many of these structure functions will vanish due to the symmetries of the governing Equations (3.22) or (3.52). Since the governing equations are symmetric with respect to change of sign of the velocities, we trivially find that all odd ordered structure functions must vanish:

$$\langle u_n^{2q+1} \rangle = 0. \tag{4.18}$$

The governing equations are also invariant under some phase rotations, $u_n \to \exp(\iota \theta_n) u_n$. The GOY model is invariant if the phases fulfil the relation

$$\theta_{n-1} + \theta_n + \theta_{n+1} = 0 \bmod 2\pi, \tag{4.19}$$

which by recursion gives, $(\theta_{3n-2}, \theta_{3n-1}, \theta_{3n}) = (\theta_1, \theta_2, -\theta_1 - \theta_2)$. Similarly, the governing equation for the Sabra model is invariant if the phases fulfil the relation

$$\theta_{n-1} + \theta_n - \theta_{n+1} = 0 \bmod 2\pi. \tag{4.20}$$

This is fulfilled if the phases form a Fibonacci sequence, $\theta_{n-1} + \theta_n = \theta_{n+1} \bmod 2\pi$. This difference between the two models gives a profoundly different behavior. The phase symmetry of the governing equation implies that the structure functions must fulfil the equation

$$\langle u_{j_1} \ldots u_{j_p} \rangle = \exp[\iota(\theta_{j_1} + \cdots + \theta_{j_p})]\langle u_{j_1} \ldots u_{j_p} \rangle, \tag{4.21}$$

where the phases fulfil the phase relations (4.19) for the GOY model or (4.20) for the Sabra model. Thus only structure functions with $\theta_{j_1} + \cdots + \theta_{j_p} = 0$ can be non-zero. This means in particular for the GOY model, that there is a possibility of long range correlations of the form $\langle u_n u^*_{n+3m} \rangle$. The modulus 3 symmetry is observed in the spectrum for the velocity in the GOY model. It is absent for the Sabra model, which makes the Sabra model more suitable for accurate measurements of spectral slopes. For the Sabra model the phases fulfil the same relations as the associated quasi-momenta (3.11), and we can conclude that only structure functions where the associated quasi-momenta sum to zero are non-zero. L'vov _et al._ (1998) argue that this makes the Sabra model superior to the GOY model.

The non-vanishing structure functions for the Sabra model can now be listed easily, observing that $\langle u_{n1} u_{n2} \ldots u_{nM} \rangle \neq 0$ implies that $\Rightarrow k_{n1} + k_{n2} + \cdots + k_{nM} = 0$. Thus we have the following 2nd, 3rd, and 4th order structure functions:

$$S_2(n) = \langle u_n u_{-n} \rangle = \langle E_n \rangle, \tag{4.22}$$

$$S_3(n) = 2\Im m \langle u_{n-1} u_n u_{-(n+1)} \rangle = \langle D_n \rangle, \tag{4.23}$$

$$S_4^{(0)}(n,m) = \langle |u_n|^2 |u_m|^2 \rangle = \langle E_n E_m \rangle, \tag{4.24}$$

$$S_4^{(1)}(n) = 2\Im m \langle u_{n-2} u_{n-1} u_{n+1} u_{-(n+2)} \rangle = -\langle D_n^{(1)} \rangle, \tag{4.25}$$

$$S_4^{(2)}(n) = 2\Im m \langle u_{n-2} u_{-n}^2 u_{n+1} \rangle = -\langle D_n^{(2)} \rangle, \tag{4.26}$$

$$S_4^{(3)}(n) = 2\Im m \langle u_{n-2} u_{n-1}^2 u_{-(n+1)} \rangle = -\langle D_n^{(3)} \rangle. \tag{4.27}$$

Here we have used short notations like $S_q(n)$ for $S_{q/2,q/2}(n;n)$, which we also use in the following when no confusion is possible.

4.4 Equivalent of the four-fifth law

Scaling exponents can be defined in complete analogy to (1.65) as

$$S_q(n) \sim k_n^{-\zeta(q)}. \tag{4.28}$$

However, it is more convenient to define the structure functions by analogy to (1.49), where the exact analytical result for a third order structure function can be seen as a benchmark. The shell model equivalent of the four-fifth law can be derived directly from conservation of energy. The exact scaling relation simply states that the mean nonlinear transfer $\Pi_n^e = \langle \tilde{\Pi}_n^e \rangle$ is independent of wave number within the inertial range in the high Reynolds number limit, $\Pi_n^e = \bar{\varepsilon}$, where $\bar{\varepsilon}$ is the mean dissipation of energy. The nonlinear flux of energy is given for the GOY model by

Equation (3.45). The equivalent formula for the Sabra model expressed in terms of the third order structure function is

$$\langle \Pi_n^e \rangle = k_n S_3(n+1) + k_{n-2} S_3(n) = \bar{\varepsilon}. \tag{4.29}$$

We return to an equivalent expression for the nonlinear flux of helicity/enstrophy in Chapter 6. From the relation (4.29) it is natural to substitute the definition of the scaling exponent (4.28) with the expression

$$\langle (\Pi_n^e)^{q/3} \rangle \sim k_n^{q/3 - \zeta(q)}. \tag{4.30}$$

From (4.29) we have $\langle \Pi_n^e \rangle = \bar{\varepsilon} \sim k_n^0$ and thus $\zeta(3) = 1$.

4.5 Problems

4.1 The real shell model (4.11) has a scaling fixed point in the case $f = 0$ and $\nu \neq 0$. Find this fixed point.

4.2 Write the list of second, third, and fourth order non-vanishing structure functions equivalent to (4.27) for the GOY model.

4.3 Verify Equation (4.29).

5

Chaotic dynamics

Chaos can be observed in simple nonlinear Hamiltonian systems. This is a dynamical system governed by Hamilton's equations where the energy is conserved, such as a physical pendulum or double pendulum. The phase space portrait of the trajectories of this kind of system can, even with few degrees of freedom, be very complicated. The phase space flow fulfils Liouville's theorem, which states that phase space volume is conserved. Another type of chaotic dynamics can arise in non-autonomous systems, like the Duffing oscillator, where a simple nonlinear system is influenced by an external periodic force. A third kind of chaotic system is nonlinear dissipative systems, such as the Lorenz (1963) model, which has only three degrees of freedom. The Lorenz model was derived from the set of ordinary differential equations describing development of wave amplitudes in the spectral representation of Rayleigh–Bernard convective flow. The Lorenz model is equivalent to a spectral truncation where only the first three wave numbers are represented. The phase space portrait of dissipative systems is different from that of Hamiltonian systems because the energy dissipation implies a shrinking of phase space volume. The dynamics of such a system is described in phase space by strange attractors. Strange attractors are sets in phase space of states u_n which are invariant with respect to the dynamical equation. This means that an initial state $u_n(0)$ belonging to the attractor will develop along a trajectory which will stay within the attractor. An initial state outside an attractor will, loosely speaking, develop along a trajectory which, after some time, will be close to the attractor. The state is thus "attracted" to the attractor. The phase space might contain more than one attractor, each attracting states within a subset of phase space. Such a subset is called the basin of attraction of the attractor. A fixed point is a trivial attractor, while strange attractors are fractal sets with noninteger dimensions lower than the dimension of the phase space. However, this description of the dynamics becomes impractical when the dimension of the system becomes large. This is the case for fluid turbulence, where the number of degrees of freedom grows with the Reynolds number (d.o.f. $\sim Re^{9/4}$).

With respect to the level of complication, the shell models are in a sense bridging the gap between the low order models and the high dimensional fluid turbulence descriptions. It is therefore of interest to investigate the chaotic properties, such as the attractor dimensionality for the shell models.

5.1 The Lyapunov exponent

A chaotic dynamical system is characterized by a rapid divergence of close-by trajectories in phase space. This has strong implications for the ability to predict the future development of the system from incompletely known initial conditions. The degree of chaos is characterized by a single number, the Lyapunov exponent. Before defining the Lyapunov exponent, we will perform a linear perturbation analysis on a dynamical system.

For any dynamical system governed by a set of n first order ODEs,

$$\dot{x}_i = f_i(\mathbf{x}), \tag{5.1}$$

a trajectory $x_i(t)$ is determined from an initial condition $x_i(0) = x_{0i}$. For any value of the parameter ϵ the perturbation $\epsilon\, y_i(0)$ of the initial condition leads to a different trajectory $x_i(t) + \epsilon\, y_i(t)$ that satisfies

$$\dot{x}_i + \epsilon\, \dot{y}_i(t) = f_i(\mathbf{x} + \epsilon\, \mathbf{y}), \tag{5.2}$$

with the initial point $x_{0i} + \epsilon\, y_i(0)$. If $|\epsilon\, \mathbf{y}(0)| \ll |\mathbf{x}_0|$ and the time is close to the initial time we can expand the right side of (5.2) to linear order in ϵ. Using (5.1) we get the linear matrix differential equation

$$\dot{\mathbf{y}} = \mathbf{A}\mathbf{y}, \quad \mathbf{A} = \{\partial_j f_i(\mathbf{x}_0)\}. \tag{5.3}$$

The solution is $\mathbf{y}(t) = e^{\mathbf{A}t}\mathbf{y}(0) = \sum_{m=0}^{\infty}[(\mathbf{A}t)^m/m!]\mathbf{y}(0)$. In the particular case where the Jacobian $\mathbf{A}$ has n linearly independent eigenvectors $\mathbf{v}_1,\ldots,\mathbf{v}_n$ with corresponding eigenvalues $\lambda_1,\ldots,\lambda_n$ we have that $\mathbf{A} = \mathbf{V}\mathbf{D}\mathbf{V}^{-1}$, where $\mathbf{V} = \{v_{ij}\}$ is the matrix with the normalized eigenvectors as columns, and $\mathbf{D} = \lceil\lambda_1\ldots\lambda_n\rfloor$ is the diagonal matrix of corresponding eigenvalues. Then it directly follows that $\mathbf{y}(t) = \mathbf{V}\sum_{m=0}^{\infty}\lceil\lambda_1^m\ldots\lambda_n^m\rfloor t^m/m!\ \mathbf{V}^{-1}\mathbf{y}(0) = \mathbf{V}\lceil e^{\lambda_1 t}\ldots e^{\lambda_n t}\rfloor\mathbf{V}^{-1}\mathbf{y}(0)$. Introducing the variable $\mathbf{z}$ by the linear mapping $\mathbf{z} = \mathbf{V}^{-1}\mathbf{y}$ we have the solution in the simple form $\mathbf{z}(t) = \lceil e^{\lambda_1 t}\ldots e^{\lambda_n t}\rfloor\mathbf{z}(0)$, where $\mathbf{z}(0) = \mathbf{V}^{-1}\mathbf{y}(0)$. Thus

$$z_i(t) = e^{\lambda_i t}z_i(0). \tag{5.4}$$

This solution determines the behavior of any perturbation of the trajectory in $\mathbf{z}$-space within a small neighborhood of $\mathbf{z}_0 = \mathbf{V}^{-1}\mathbf{x}_0$, where $\mathbf{z}_0 \neq \mathbf{z}(0) = \mathbf{V}^{-1}\mathbf{y}(0)$.

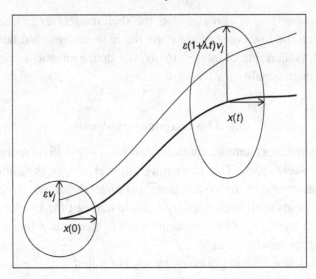

Figure 5.1 The linear development of a neighborhood of the trajectory $\mathbf{x}(t)$ near $t = t_0$. The principal axes of the ellipsoid are determined by the eigenvalues of the Jacobian matrix.

Note that the $\mathbf{z}$-space is just an affine map of the $\mathbf{x}$-space. In particular the $\mathbf{z}$-space is obtained as a rotation of the $\mathbf{x}$-space if the Jacobian $\mathbf{A}$ is symmetric, implying that $\mathbf{V}$ exists and is orthogonal so that $\mathbf{V}^{-1} = \mathbf{V}^\mathsf{T}$. This situation is present if the force vector f_i is defined as the gradient of a potential field.

A sphere in $\mathbf{z}$-space centered at $\mathbf{z}_0$ with radius ϵ will evolve to an ellipsoid centered at $\mathbf{V}^{-1}\mathbf{x}(t)$ with principal axes along the unit vectors of the $\mathbf{z}$-space and the lengths of the axes $\epsilon\,\mathrm{e}^{\lambda_i t} \approx \epsilon\,(1 + \lambda_i t)$. This is directly seen from the equation of the ellipsoid considered in the translated coordinate system that has its origin at the point $\mathbf{V}^{-1}\mathbf{x}(t)$: $\mathbf{z}(0)^\mathsf{T}\mathbf{z}(0) = \mathbf{z}^\mathsf{T}\lceil \mathrm{e}^{-2\lambda_1 t}\ldots\mathrm{e}^{-2\lambda_n t}\rfloor\mathbf{z} = \epsilon^2$, see Figure 5.1. In the $\mathbf{x}$-space, these principal axes correspond to conjugate axes along the eigenvectors $\mathbf{v}_i$.

In the $\mathbf{z}$-space, the linearized differential equation reads $\dot{\mathbf{z}} = \lceil \mathrm{e}^{\lambda_1 t}\ldots\mathrm{e}^{\lambda_n t}\rfloor\mathbf{z}$. Thus the vector $\lceil \mathrm{e}^{\lambda_1 t}\ldots\mathrm{e}^{\lambda_n t}\rfloor\mathbf{z}$ is tangential to the trajectory passing the point $\mathbf{z}$. Setting $\mathbf{z}$ to $\lceil \mathrm{e}^{\lambda_1 t}\ldots\mathrm{e}^{\lambda_n t}\rfloor\mathbf{z}_\mathrm{e}$ in the ellipsoid equation gives the equation

$$\mathbf{z}_\mathrm{e}^\mathsf{T}\lceil \mathrm{e}^{\lambda_1 t}\ldots\mathrm{e}^{\lambda_n t}\rfloor\lceil \mathrm{e}^{-2\lambda_1 t}\ldots\mathrm{e}^{-2\lambda_n t}\rfloor\lceil \mathrm{e}^{\lambda_1 t}\ldots\mathrm{e}^{\lambda_n t}\rfloor\mathbf{z}_\mathrm{e} = \mathbf{z}_\mathrm{e}\mathbf{z}_\mathrm{e}^\mathsf{T} = \epsilon^2. \qquad (5.5)$$

Thus there is no perturbation in the direction of the trajectory in the $\mathbf{z}$-space. This is also the case in the $\mathbf{x}$-space if the Jacobian is symmetric, but otherwise it is not necessarily so. The ellipsoid is contracting in the directions with $\lambda_j < 0$, unchanged for $\lambda_j = 0$, and expanding in the directions where $\lambda_j > 0$. The spectrum of eigenvalues depends on the point $\mathbf{x}_0$ where the linear expansion was made. In order to characterize the dynamics of the system independently of x_0 we want to follow the trajectories all over the attractor. Thus we want to construct a "mean" Jacobian matrix from which we can calculate the eigenvalue spectrum. We do that by considering a discrete map.

The governing equation (5.1) can be approximated by a difference equation:

$$x_i(t + \Delta t) = x_i(t) + f_i[\mathbf{x}(t)]\Delta t \equiv g_i[\mathbf{x}(t)]. \tag{5.6}$$

This defines the trajectory as a sequence, $\mathbf{x}_0, \mathbf{x}_1 = \mathbf{x}(t + \Delta t), \ldots, \mathbf{x}_n = \mathbf{x}(t + n\Delta t), \ldots$, and the difference equation is a map, $\mathbf{x}_{n+1} = \mathbf{g}(\mathbf{x}_n)$. A linear perturbation will evolve as $\epsilon_{n+1} = J_n \epsilon_n$, where $J_n = \partial g_n$ is the Jacobian (suppressing matrix indices). The development of a perturbation after m steps is then $\epsilon_{n+m} = J_n J_{n+1} \ldots J_{n+m} \epsilon_n$. The eigenvalue spectrum of the matrix,

$$J = \lim_{m \to \infty} [J_n J_{n+1} \ldots J_{n+m}]^{1/m}, \tag{5.7}$$

is called the *Lyapunov numbers*, and the eigenvectors are the *Lyapunov vectors*. The algebraic operations such as taking the power $1/m$ of a matrix make sense by expressing the matrix on the eigenvector basis as a diagonal matrix and applying the operation to the eigenvalues in the same way as was done for the linear analysis in (5.4). The Lyapunov numbers are usually labeled in descending order $\lambda_1 > \lambda_2 > \ldots$ The limit $m \to \infty$ ensures that the spectrum becomes independent of the initial state $\mathbf{x}_n$ and that the whole attractor is sampled. The eigenvectors will then evolve as $v_m = \lambda^m v_0 = \exp(m \log \lambda) v_0$. With Δt constant, m signifies time and the exponential development of initial perturbations is seen. The exponents $\sigma_i = \log \lambda_i$ are the *Lyapunov exponents*. Returning to the differential equation (5.1), using the short notation $\epsilon_i(t) = \epsilon x_i'(t)$, the Lyapunov exponents are defined as

$$\sigma_i = \lim_{t \to \infty} \frac{1}{t} \log\left(\frac{\epsilon_i(t)}{\epsilon_i(0)}\right), \tag{5.8}$$

where ϵ_i is the length of a perturbation in the direction of the ith eigenvector of the Jacobian matrix. The limit $t \to \infty$ corresponds to the limit $m \to \infty$ in the discrete case. The validity of the linear Equation (5.3) is ensured by having the perturbation $\epsilon_i(0)$ sufficiently small. The ratio R of the volume of the perturbation ellipsoid to the volume of the initial sphere is

$$R = \Pi_i \lambda_i = \exp\left(\sum_i \sigma_i\right). \tag{5.9}$$

For a Hamiltonian system the phase space volume is conserved and thus $\sum_i \sigma_i = 0$. For a dissipative system the phase space volume is shrinking, and $\sum_i \sigma_i < 0$. There will always be at least one Lyapunov exponent which is zero, namely the one corresponding to the direction of the unperturbed trajectory. The system will be chaotic if at least one Lyapunov exponent is positive, corresponding to a direction where a perturbation grows exponentially. The degree of predictability is thus determined by the largest of the Lyapunov exponents σ_1, which is called the maximum Lyapunov exponent, or just the Lyapunov exponent. The Lyapunov exponent is easily accessible in a numerical simulation. For a simulation sufficiently long, which means that the attractor has been swept sufficiently densely for the statistics to be stable, we insert infinitesimal perturbations and follow the perturbed trajectories until their distance in phase space from the unperturbed trajectory is of the order of the diameter of the attractor. The Lyapunov exponent is the average exponential growth rate of the distances as symbolically depicted in Figure 5.2.

The distances between perturbed trajectories for a simulation of the Sabra shell model with $N = 20$ shells are shown in Figure 5.3. The distances grow approximately exponentially until they are of the same order of magnitude as the size of the attractor. From the average slope in the lin-log plot the maximum Lyapunov exponent is measured, as indicated by the thick line, which in this case corresponds to a Lyapunov exponent of 11.5 time units^{-1}.

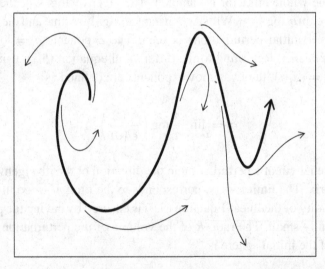

Figure 5.2 The thick curve represents a trajectory in phase space, the thin curves are perturbed trajectories. The Lyapunov exponent is determined from the mean exponential separation in time between the unperturbed (thick) trajectory and the perturbed (thin) trajectories.

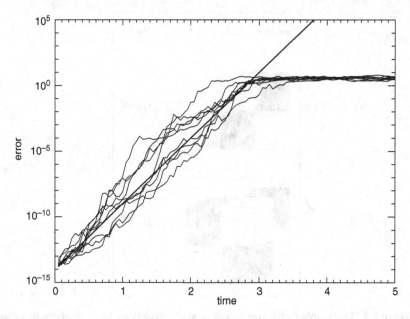

Figure 5.3 The growth of distances between ten trajectories in phase space where a small random perturbation between the two introduces a distance of the order 10^{-14}. The size of the attractor is of the order unity so the distances after about three time units are of the same order as the distances between any two trajectories. The distance in phase space is defined as the Euclidean norm.

5.2 The attractor dimension

Since the phase space volume is shrinking with time in a dissipative system, the volume of the attractor must be zero in the N dimensional phase space; N is the number of degrees of freedom. The attractor is thus not characterized by its volume but by its dimensionality. A fixed point has dimension zero, while a chaotic system usually is characterized by a non-integral attractor dimension. Such an attractor is called a strange attractor. There are various formulations and definitions of the dimension of the attractor in a chaotic system. The most straightforward definition is the Hausdorff dimension, D_H. The Hausdorff dimension of a set $A \in R^N$ is defined geometrically (Mandelbrot, 1977). For a given small number ϵ the minimum number of open balls of radius ϵ that can be chosen to cover the set A is $N(\epsilon)$. This number of points scales with ϵ as

$$N(\epsilon) \sim \epsilon^{D_H}, \tag{5.10}$$

in the limit $\epsilon \to 0$. In order to ensure that $N(\epsilon)$ is finite, A must be compact, however, if this is not the case we can perform the analysis on all compact subsets

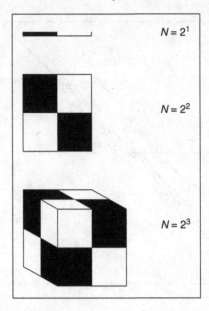

Figure 5.4 The number of boxes of linear size l necessary to cover a compact set of dimension D_H scales with l as $N(l) \sim l^{D_H}$. D_H is the Hausdorff dimension of the set.

of A instead. It is seen from Figure 5.4 that this definition reproduces the dimension of integral dimensional spaces.

For a dynamical system, however, this is an impractical definition for calculating the dimension. A more practical definition was given by Grassberger and Procaccia (1983). A chaotic dynamical system will most often be explored by generating a sample from a trajectory, $x_i = x(t_i)$, $i = 1, \ldots, M$. The M points will be situated on the attractor. If the sampling intervals $t_{i+1} - t_i$ are longer than the dynamical correlation time, which we can define as the inverse of the Lyapunov exponent, then pairs of points (x_i, x_j) will only be spatially correlated through the geometry of the attractor and the recurrence of points within regions of the attractor. A measure of this is the correlation integral defined as

$$c(l) = \frac{1}{M(M-1)} \sum_{ij} 1_{[0,l]}(|x_i - x_j|), \qquad (5.11)$$

where $1_{[0,l]}$ is the indicator function. So $c(l)$ is the average number of points with distances less than l from any given point. It is denoted the correlation integral since it is the integral of the two point correlation function from 0 to l. If points are situated evenly across the attractor, the correlation integral must scale with l as

$$c(l) \sim l^{\nu}, \tag{5.12}$$

where ν is called the correlation or embedding dimension.

A third useful concept is the information dimension. This is inspired by the definition of entropy in statistical mechanics. If we consider a coverage of phase space by boxes of linear size l_0, corresponding to a coarse graining of the system, we can define entropy as

$$S(l_0) = -\langle \log p \rangle = -\sum_i p_i \log p_i, \tag{5.13}$$

where $p_i = n_i/M$ is the relative frequency of visiting box i. This information entropy can be considered as a measure of the number of "micro-states," corresponding to actual trajectories in phase space, giving the same coarse grained "macro-state." The information entropy signifies the amount of information required to determine in which macro-state the system is. If we refine the coarse graining to linear scale $l' = l_0/l$, then the attractor will be covered by l^{D_H} boxes. Within each of the original boxes the relative frequency of visiting a given smaller box within box i is $p_i l^{-D_H}$, and we get

$$S(l') = -\sum_i \sum_{j=1}^{l^{D_H}} p_i l^{-D_H} \log(p_i l^{-D_H}) \tag{5.14}$$

$$= -\sum_i p_i (\log p_i - D_H \log l) = S(l_0) + D_H \log l.$$

By choosing $l_0 = 1$ and writing $S(1) = S_0$ we can rewrite this as

$$S(l) = S_0 - \sigma \log l, \tag{5.15}$$

where $\sigma = D_H$ is called the information dimension. The higher the information dimension, the more information is required to determine the state of the system as the resolution is increased, just as the number of coordinates increases when specifying the position in higher dimensional spaces. In the derivation above we used the assumption that all sub-boxes had the same relative frequency of being visited, leading to the equality $\sigma = D_H$. If this is not the case, we can only establish the inequality

$$\sum_j p_i^j \log p_i^j \le p_i \log p_i, \tag{5.16}$$

which follows from the convexity of the function $-x \log x$. This in turn gives the inequality $\sigma \le D_H$. Likewise the correlation dimension is only upwardly bound by the Hausdorff dimension and can be smaller in the case when the attractor is not uniformly visited. Grassberger and Procaccia (1983) established the inequalities

$$\nu \le \sigma \le D_H, \tag{5.17}$$

where the equalities are valid if the attractor is visited uniformly. The metric dimension of the attractor is connected to the dynamics through the concept of information. In a chaotic system, the information of the initial state is lost after some (de-)correlation time and after a while the information needed to specify the state of the system is related to the information dimension of the attractor. The rate of loss of information about the initial state is related to the separation of trajectories. An infinitesimal sphere of radius ϵ in phase space will evolve in a time interval δt to an ellipsoid with major axes along the Lyapunov vectors and lengths $\epsilon \exp(\lambda_i \delta t)$, where λ_i is the ith Lyapunov exponent. In order to cover the ellipsoid with spheres of radius ϵ we need $\sim \exp(\sum_{\lambda_i>0} \lambda_i \delta t)$ spheres. So we lose information in the direction of positive Lyapunov exponents while there is no loss in the direction of negative Lyapunov exponents; see Figure 5.5. This means that a point

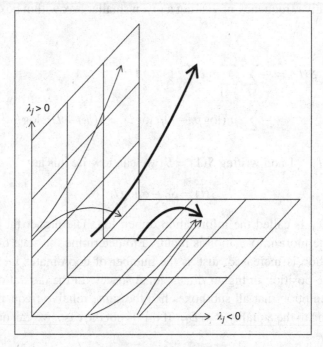

Figure 5.5 Information on the position in phase space is lost in the direction of positive Lyapunov exponents while it is retained in the direction of negative Lyapunov exponents. The thick curves are two nearby trajectories in phase space, while the thin curves are the projections of the trajectories in directions i and j, respectively.

inside the original sphere after time δt can be in either of $\exp\sum_{\lambda_i>0}\lambda_i\delta t$ spheres, and the rate of change of information, being proportional to the logarithm of the number of possible spheres, is the sum of the positive Lyapunov exponents. This sum is called the *Kolmogorov–Sinai entropy* of a dynamical system. Along these lines, Kaplan and Yorke (1979) conjectured that the attractor dimension could be calculated as

$$D_{KY} = j + \frac{\lambda_1 + \lambda_2 + \cdots + \lambda_j}{|\lambda_{j+1}|}, \tag{5.18}$$

where the Lyapunov exponents are ordered in descending order and j is the smallest index such that $\lambda_1 + \lambda_2 + \cdots + \lambda_j \geq 0$. The attractor dimension is larger than or equal to j; an infinitesimal hyper-volume of dimension j spanned by the first j eigenvectors will increase or stay constant since $\sum_{i\leq j}\lambda_i \geq 0$. On the other hand, the attractor dimension is smaller than $j+1$; $j+1$ dimensional infinitesimal hyper-volumes spanned by the first $j+1$ eigenvectors will shrink since $\sum_{i\leq j+1}\lambda_i < 0$. The non-integral part of the dimension is simply the linear interpolation to zero of the function $\sum_{i=1}^{k}\lambda_i$ as a function of k. It is not yet clear under what circumstances, if any, the conjecture is rigorously correct. Grassberger and Procaccia (1983) argue that the Kaplan–Yorke dimension is an upper bound for the correlation dimension.

5.3 The attractor for the shell model

A flow field is high dimensional and the number of degrees of freedom scales with the Reynolds number. In 3D turbulence, we will have to resolve scales from the integral scale L to the dissipative scale η, which in 3D requires $(L/\eta)^3$ degrees of freedom. With $\eta \sim Re^{-3/4}$ we have the effective number of degrees of freedom growing as $Re^{9/4}$. For shell models the effective number of degrees of freedom is simply (twice) the number of shells from the integral to the dissipation scale. The Reynolds number is defined as $Re = U_1 k_1^{-1}/v$, where U_1 is a typical velocity at shell $n = 1$ (integral scale). The dissipation scale k_E, index E indicating "energy dissipation," is obtained in complete analogy to (1.11). From energy conservation we have $v k_E^2 U_E^2 \sim \bar{\varepsilon} \sim k_n U_n^3$, for $1 \leq n < n_E$. This leads to $U_n \sim \bar{\varepsilon}^{1/3} k_n^{-1/3}$ and for $n = n_E$ we get $k_E \sim (\bar{\varepsilon}/v^3)^{1/4}$, thus finally $k_E \sim Re^{3/4}$. The number of degrees of freedom follows from $k_E = k_0 \lambda^{n_E}$, implying that $n_E \sim \log k_E \sim (3/4)\log Re$, which is the upper limit for the attractor dimension. The relevant intensive measure of the attractor dimension is then the ratio of the dimension of phase space expressed in terms of the Reynolds number and the attractor dimension. Similarly, the Lyapunov spectrum should be normalized with respect to the attractor dimension

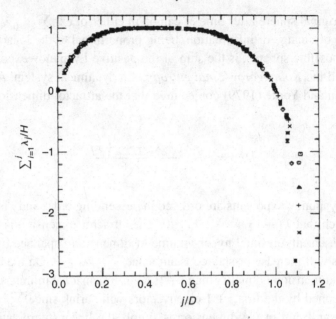

Figure 5.6 The Lyapunov spectrum represented as $\sum_{i=1}^{j} \lambda_j / H$, where H is the Kolmogorov–Sinai entropy, versus j/D, where D is the Kaplan–Yorke dimension. The spectrum is obtained from numerical simulations of the GOY model (Yamada & Okhitani, 1988b).

or the maximum Lyapunov exponent. The Lyapunov spectrum has been calculated for the GOY model (Yamada & Okhitani, 1988b). A nice treatment can be found in Bohr *et al.* (1998). Figure 5.6 shows the spectrum $\sum_{i=1}^{j} \lambda_j / H$ versus j/D, where H is the Kolmogorov–Sinai entropy and D is the Kaplan–Yorke dimension. The spectrum is independent of Reynolds number and seems to converge to a continuous curve for $\mathsf{R}e \to \infty$. The curve indicates a finite density of vanishing Lyapunov exponents. The time average of the squared components of the Lyapunov vectors is shown in Figure 5.7. Here the first axis (i) denotes the shell number while the second axis (j) denotes the eigenvector corresponding to eigenvalue λ_j. The figure shows that the eigenvectors corresponding to positive Lyapunov exponents (approximately the first 10 vectors) are concentrated in the inertial range (shell numbers 10–30) with the most unstable modes at the largest shell numbers. The localization of eigenmodes in the shell space was called "Lyapunov–Fourier correspondence" by Yamada and Okhitani (1988b). The eigenvectors (10–30) show a weak Lyapunov–Fourier correspondence and are represented by the flat part of the spectrum in Figure 5.6 with vanishing Lyapunov exponents. The eigenvectors (30–45) with strong Lyapunov–Fourier correspondence are the contracting dissipative modes.

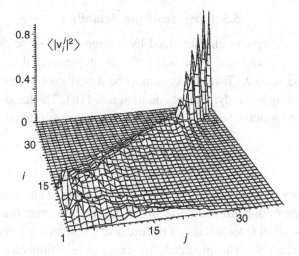

Figure 5.7 A "bird's eye" view of the time average of the square of the components of the Lyapunov vectors, $\langle |v_i^j|^2 \rangle$, where i is the shell number (Fourier index), λ_j is the corresponding Lyapunov exponent, and j is the Lyapunov index (Yamada & Okhitani, 1988b).

5.4 Predictability

Predicting the future development of a chaotic system from an imperfectly known initial state, such as in weather forecasting, is an important practical problem. If the system has a positive Lyapunov exponent, $\lambda > 0$, a small error $\epsilon = x_{\text{true}} - x_{\text{initial}}$ will, on average, evolve as $\epsilon(t) = \epsilon(0) \exp \lambda t$. This means that if only an error smaller than E is tolerable, then the true initial state must be determined with an accuracy $\epsilon = E \exp(-\lambda T)$ in order to predict the development of the system a time T into the future. This makes forecasting beyond times of the order λ^{-1} fundamentally impossible. We may then define the inverse of the Lyapunov exponent to be the predictability time. In the simulation of the Sabra model (Figure 5.3), we calculated $\lambda^{-1} \sim 0.09$ time units. When comparing this to the time development of the velocities in Figure 4.2, it is obvious that the predictability time is associated with the variations at the fastest timescales near the dissipative range. In the atmosphere, the Lyapunov exponent will be related to variations which are irrelevant in connection with weather forecasts. Likewise, the development of a cumulus cloud is predictable at perhaps 15 minute timescales, while the timescale relevant for development of cyclonic disturbances is of the order of days, which is the timescale relevant for weather predictions. At these timescales the cumulus convection is effectively random noise acting as a source of error in the cyclonic development.

5.5 Large scale predictability

Consider a chaotic system characterized by a large range of scales, such as the atmosphere. We are interested in predicting the development of the system on scales larger than some scale L. Take the system to be described by some field variable $\mathbf{u}(\mathbf{x}, t)$, governed by some dynamical equation $\dot{\mathbf{u}} = \mathbf{f}(\mathbf{u})$. The large scales are then described by $\mathbf{u}_{\geq L}$, which is some coarse-grained or low-pass filtered representation:

$$\mathbf{u}_{\geq L}(\mathbf{x}, t) = \int_{k \leq L^{-1}} e^{\iota \mathbf{k}\mathbf{x}} \mathbf{x}(\mathbf{k}) d\mathbf{k}. \tag{5.19}$$

The predictability time will be of the order of the timescale characteristic for variations at the scale L. In a turbulent flow this is the eddy turnover time. For synoptic scale weather, skillful forecasts are of the order of days, which is the timescale for baroclinic wave activity. This predictability time can be estimated by assuming the dynamics described by an effective large scale dynamics, $\dot{\mathbf{u}}_{>L} = \mathbf{f}_{\text{eff}}(\mathbf{u}_{>L})$. The predictability time would then be determined from the Lyapunov exponent associated with $\mathbf{f}_{\text{eff}}$. In the case of shell models, the large scales are described by the shell variables, u_m, for $m < n$ where $k_n = L^{-1}$. The predictability time for the large scales would then be of the order $\tau_n = (k_n u_n)^{-1} \sim k_n^{-2/3}$, which is much longer than the inverse of the Lyapunov exponent for the full system. The limited predictability of the development at large scales defined in this way arises from the inherent chaos at the large scales.

More fundamentally, the limited predictability is caused by imperfectly known initial conditions. Imagine an ideal observational system, where the the initial state is perfectly known on all scales larger than some scale ϵ. This could be by, say, observing the state in every grid-point in a mesh with a lattice spacing ϵ. Then the timescale for predictability on scale L will be governed by the time it takes to propagate the error at scale ϵ to the scale L. In a turbulent flow, the timescale for propagating an error from scale $l/2$ to scale l, or vice versa, will be of the order of the deformation time or eddy turnover time at scale l. The predictability time is thus a sum of eddy turnover times covering the range $\epsilon < l < L$. If this predictability time remains finite as $\epsilon \to 0$, this will be a fundamental barrier for long range forecasting. In particular, for the shell models, the predictability time at shell n will be of the order

$$T_n \sim \sum_{m=n}^{\infty} \frac{1}{k_m u_m} \sim k_n^{\alpha-1} \sum_{m=0}^{\infty} \lambda^{m(\alpha-1)} = k_n^{\alpha-1} \frac{1}{1 - \lambda^{\alpha-1}}, \tag{5.20}$$

where we have assumed the scaling $u_m \sim k_m^{-\alpha}$. The flow is then fundamentally unpredictable if $T_n < \infty$ which is the case for $\alpha < 1$. The energy spectrum is

$E_m \sim k_m^{-2\alpha}$, so the more energy the small scales contain the faster the error propagation and the shorter the predictability time. For the 3D case, $\alpha = 1/3$, and thus the predictability time is finite, while for the 2D case, $\alpha = 1$ and thus the potential for long range forecasting exists.

The considerations presented here are based on a seminal paper by Lorenz (1969) in which he derives the linearized dynamics of the error field from the vorticity equation. This leads to the same kind of up scale error propagation as described here. Lorenz calculated the predictability based on the $k^{-5/3}$ 3D turbulence energy spectrum. This has much more energy in the small scales than the atmosphere has, such that he underestimated the predictability time in the synoptic scale atmosphere.

The predictability time based on error propagation from small scales leads to the same scaling with k_n as the consideration based on the effective large scale dynamics. So even though the time for errors at infinitesimally small scales to propagate to the large scales could be infinite, the predictability will still be limited due to the chaotic dynamics at large scales with critical dependence on initial conditions.

5.6 The finite size Lyapunov exponent

The observation that predictability in the large scales is independent of the Lyapunov exponent has been quantified in the *finite size Lyapunov exponent* (FSLE) (Aurell *et al.*, 1996a, 1996b). The qualification "finite size" refers to an error of finite size in the large scales initial conditions in contrast to an infinitesimal error. The idea is that the relevant timescale of predictability should be associated with the timescale of disturbances at the scale of interest. Consider the case of weather predictions. The Lyapunov exponent for the atmospheric flow, which we for now will denote the maximum Lyapunov exponent λ_{max}, is associated with small wind swirls, or in the free atmosphere, perhaps with convective developments. So we have $\lambda_{max} \geq 1/$minutes. The relevant scales of the weather are baroclinic waves of wavelength of the order of 1000 km. These are predictable for a few days, which corresponds to the eddy turnover time of these waves or eddies. Thus the maximum Lyapunov exponent is irrelevant for weather predictions.

Instead the FSLE is constructed to reflect the scale of interest. Following Aurell *et al.* (1996a), the growth of an error in the velocity field can be measured in the Euclidean norm

$$\delta u(t)^2 = ||\delta u||^2 \equiv \int [u'(\mathbf{x}, t) - u(\mathbf{x}, t)]^2 d\mathbf{x} = 2 \int E'(\mathbf{x}, t) d\mathbf{x}, \qquad (5.21)$$

where we have defined the "error energy field" $E'(\mathbf{x}, t)$. The velocity error $\delta u(t)^2$ will be a measure of the error at the largest scale of disturbance. This follows from

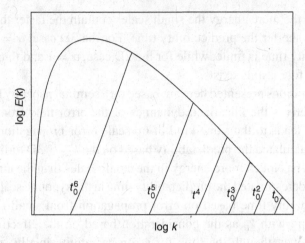

Figure 5.8 The error field is the difference between two fields initially separated by an infinitesimal error. The error energy spectrum develops with time to cover the whole energy spectrum (top curve). The spectrum of the error field and the energy spectrum coincide to the right of the intersection. Only scales to the left of the error fields are predictable for times longer than the time label associated with the error curve. The graph is schematic and adopted from Lorenz (1969).

the K41 expectation of the spectral shape of the error energy, $E'(k) \sim k^{-5/3}$ for $k > 1/L(t)$, where $L(t)$ is the largest scale the disturbance has reached at time t; see Figure 5.8. The error energy spectrum $E'(k)$, suppressing the dependence on time, is obtained from the Fourier transform $E(\mathbf{k}, t)$ of the error energy field by integrating out the angular coordinates of the wave vector $\mathbf{k}$. Thus, using Parseval's equality, we have $\delta u(t)^2 \sim \int_{1/L(t)}^{k_E} k^{-5/3} dk \sim L(t)^{2/3}$. Given an initial error $\delta u(0)^2 = \delta_0^2$, the numerical size of this initial error will be associated with some spatial scale L. We can measure the time T it takes for this error to grow to some threshold value Δ. Natural choices for this could be $\Delta = 2\delta_0$ (doubling time) or $\Delta = e\delta_0$ (e-folding time, where e is the base of the natural logarithm). Defining the ratio $r = \Delta/\delta_0$, we eliminate Δ. The time is then given as.

$$T_r(\delta_0) = \min\{t \geq 0 \,|\, \delta u(t') \leq r\delta_0 \text{ for } t' < t\}. \tag{5.22}$$

When the disturbance δ_0 is infinitesimal, linear analysis gives

$$\Delta(t)/\delta_0 \approx \exp \lambda_{\max} t. \tag{5.23}$$

Analogous to this, the FSLE is defined as

$$\lambda(\delta_0) = \left\langle \frac{1}{T(\delta_0)} \right\rangle \log r, \tag{5.24}$$

where the brackets denotes the temporal average. This average is calculated by the following numerical procedure: a trajectory $\mathbf{u}(t)$ in phase space is calculated from the governing equation. Then a perturbed trajectory $\mathbf{u}'(t)$ is calculated by initiating from $\mathbf{u}'(0) = \mathbf{u}(0) + \delta_0 \epsilon / ||\epsilon||$, where δ_0 is some random perturbation (error) in the direction $\epsilon / ||\epsilon||$. The dynamics is then integrated until the error has grown to the size $r\delta_0$. At this time τ_1 the perturbed field $\mathbf{u}'$ is rescaled $\mathbf{u}' \rightarrow \mathbf{u} + \delta_0(\mathbf{u}' - \mathbf{u})/||\mathbf{u}' - \mathbf{u}||$. The dynamics is then integrated until time τ_2 where the error again is $r\delta_0$ and so on. The error-mean doubling time is then denoted by a suffix $\langle\rangle_e$ and defined as

$$\langle T \rangle_e = \frac{1}{N} \sum_{i=1}^{N} (\tau_i - \tau_{i-1}) = \frac{1}{N} \sum_{i=1}^{N} T_i, \qquad (5.25)$$

where $T_i = \tau_i - \tau_{i-1}$. The error-mean is not the same as the temporal mean, since the doubling times are not weighted with the time they last. The relation between the two means is obtained by taking the temporal mean of $1/T_i$:

$$\left\langle \frac{1}{T} \right\rangle = \frac{1}{\tau_N} \int_0^{\tau_N} \frac{1}{T_i} dt = \frac{1}{\tau_N} \sum_{i=1}^{N} \frac{1}{T_i} T_i = \frac{N}{\tau_N} = \frac{1}{\langle T \rangle_e}. \qquad (5.26)$$

From this it follows that the maximum Lyapunov exponent is obtained in the limit $\lambda_{\delta \to 0} = \lambda_{max}$.

For a high Reynolds number flow we can apply K41 theory. The only timescale relevant at a spatial scale l is the eddy turnover time $\tau(l) = l/\delta u(l)$. The FSLE $\lambda(\delta u)$ for a perturbation of the order $\delta = \delta u(l)$ then becomes

$$\lambda(\delta) \sim \tau(l)^{-1} \sim \delta/\delta^3 = \delta^{-2}, \qquad (5.27)$$

where we use the K41 relation $\delta u(l) \sim l^{1/3}$. This should hold for δ large in comparison to $U(\eta/L)^{1/3}$. When δ becomes of the order of the fluctuations at the Kolmogorov scale we should expect an exponential growth of errors, and thus $\lambda(\delta) \approx \lambda_{max}$ independent of δ. The behavior of the FSLE is shown in Figure 5.9 for the case of the GOY shell model.

5.7 Limited predictability of the small scales

A different nature of limited predictability due to error growth at the small scales was argued by Ruelle (1979). The idea is that thermal fluctuations representing the chaotic motions of the molecules in the fluid would propagate to the Kolmogorov scale and thus effectively randomize the velocity field at this scale. This in turn would result in fundamentally limited predictability at the large scales. Ruelle suggested that the predictability time at the large scales in the flow will be dominated by

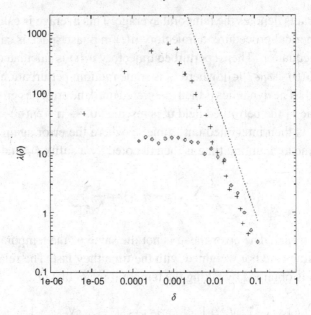

Figure 5.9 The finite size Lyapunov exponent $\lambda(\delta)$ as a function of the non-infinitesimal initial perturbation δ. For $\delta \to 0$, the maximum Lyapunov exponent emerges, while for δ corresponding to typical velocity fluctuations in the inertial range the K41 prediction is $\lambda(\delta) \sim \delta^{-2}$, as indicated by the dashed line. The diamonds are calculated for the GOY shell model (Aurell *et al.*, 1996a).

this propagation of thermal fluctuations. The argument combines chaos and K41 theories; the error in the velocity field originating from the thermal fluctuations grows exponentially from the microscopic scale

$$\Delta(t) = \Delta(0)e^{\lambda t}. \tag{5.28}$$

Here λ is the Lyapunov exponent that represents the fastest error growth rate in the flow. This growth rate is associated with the timescale obtained for the smallest scales in the flow

$$\lambda \sim \delta u(\eta)/\eta \sim v/\eta^2 \sim Re^{-1} Re^{3/2} = Re^{1/2}, \tag{5.29}$$

where $\eta = (v^3/\bar{\varepsilon})^{1/4} \sim Re^{-3/4}$ is the Kolmogorov scale. A microscopic error in the velocity is noticeable when it becomes of the order of the velocity differences at the Kolmogorov scale, $\Delta(t) \sim \delta u(\eta) = v/\eta$. Now the timescale t can be obtained from (5.28) and (5.29) if we can estimate $\Delta(0)$. This must be related to the microscopic fluctuations in molecular velocities. A volume of fluid with the Kolmogorov scale as linear size is constituted of $N\eta^3$ molecules, where N is the number of molecules per unit volume (of the order of Avogadro's number). The mean velocity of the molecules is of the order $c = \sqrt{kT}$, where k is Boltzmann's constant and T is the temperature. The fluctuations will be of the order

$$\Delta(0) \sim \sqrt{kT}/\sqrt{N\eta^3}, \qquad (5.30)$$

where the last square root comes from averaging. Using $\Delta(t) \sim \nu/\eta$ and inserting (5.30) into (5.28) we obtain

$$t \sim \lambda^{-1} \log(\nu\sqrt{N\eta/kT}). \qquad (5.31)$$

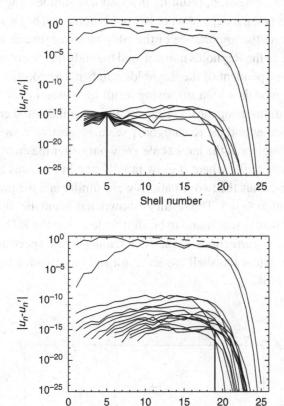

Figure 5.10 The growth of an infinitesimal error in the Sabra model. In the top panel the error is introduced in shell number 5 (vertical bar). In the lower panel the error is introduced at shell number 19 (vertical bar). The curves represent the error velocity fields $|u_n - u'_n|$ for an exponentially increasing sequence of times. The lowermost curves are error spectra after a short time, while the topmost curve is the error spectrum after a long time. In both cases the inverse error propagation begins when the error reaches the Kolmogorov scale. The dashed lines correspond to the Kolmogorov scaling $u \sim k^{-1/3}$. (Adapted from Crisanti et al. (1993b) where the same simulation was presented for the GOY model.)

This shows that the predictability time is the eddy turnover time at the Kolmogorov scale except for a logarithmic factor, which is not very sensitive to its argument. The logarithmic dependence of the microscopic properties of the fluid is a consequence of the assumption (5.28). However, there is no reason to believe that the assumption of exponential separation of trajectories is valid all the way from the microscopic scale to the Kolmogorov scale. One would rather expect a growth rate proportional to $\sqrt{t}$ as a result of many stochastic contributions from molecular collisions. However, the scaling argument leading to the $Re^{1/2}$ dependence still holds. The argument presented above is noteworthy because it is in contrast to the result from error propagation, resulting in a quotient sum leading to predictability at the large scales being independent of Reynolds number. The two different results simply demonstrate the importance of the eddy turnover time at the Kolmogorov scale, depending on the Reynolds number, and the eddy turnover time at the integral scale, which is independent of the Reynolds number. The conflict was resolved in the case of shell models with a surprising result by Crisanti *et al.* (1993a). In two numerical simulations using the GOY model, infinitesimal initial errors of the same size at the large scales and the Kolmogorov scale, respectively, were introduced. It was found that the error at the large scale grew only after the error had propagated to the Kolmogorov scale, where it grew rapidly and in turn was back propagated to the large scales. Thus the two situations were similar and the predictability time grows as Re^{α} with $\alpha \approx 0.5$. The result is shown diagrammatically in Figure 5.10. The scaling exponent α was found to be slightly less than the K41 value of 1/2 due to the intermittency corrections to the K41 scaling of the spectrum. It is an open question to what extent the shell model result can be extended to turbulent flows governed by the NSE.

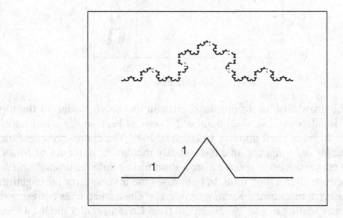

Figure 5.11 The Koch curve.

5.8 Problems

5.1 Fractal dimension. Determine the fractal dimension of the object shown in the top of Figure 5.11, which is one of the first fractals described, called the Koch curve. It is obtained by iteratively substituting all line segments with a correspondingly scaled version of the object at the bottom. This is one third part of the Koch star, of which the first iteration is an equilateral triangle and the second iteration is the star of David.

5.2 Non-Hamiltonian dynamics. Consider the dynamical system:

$$\dot{x} = xy,$$
$$\dot{y} = -x^2.$$

Show that this system conserves energy, $E = x^2 + y^2$.
Sketch the phase space flow for any initial state in the xy-plane.
Does this system fulfil Liouville's theorem?
Is this a Hamiltonian system? If not, explain why.

5.3 The logistic map has a surprisingly rich and complex behavior, which can be seen as an analog to the much more complex behavior of the NSE. The logistic map is given by:

$$x_{x+1} = f(x_n) = rx_n(1 - x_n),$$

where $0 \leq x_n \leq 1$, and $0 < r \leq 4$ is a parameter.
Plot x_{n+1} versus x_n in a figure and determine for which values of r the map has a fixed point ($x = f(x)$). We shall not be concerned with the trivial fixed point $x = 0$.
By performing a linear stability analysis, determine for which values of r the fixed point is stable. Do that by examining the evolution of the map beginning from $x_1 = x_0 + \epsilon$, where x_0 is the fixed point and $\epsilon \ll 1$.
Examine what happens when the fixed point becomes unstable (as r is increased) by investigating the map $f^2(x_n) = f(f(x_n))$.
This approach can be continued by examining $f^4, \ldots, f^n$, but the algebra rapidly becomes very cumbersome. What happens is called the "period-doubling route to chaos," and if the fixed points are plotted as a function of r the so-called Feigenbaum tree emerges; this is a beautiful fractal object. Here we shall not continue further, but if you want to learn more a good starting point is http://mathworld.wolfram.com/LogisticMap.html

5.4 The Lorenz attractor. The Lorenz model (Lorenz, 1963) is a set of three coupled nonlinear ordinary differential equations. It is derived from a highly truncated spectral representation of convection rolls in the atmospheric flow originally derived by Saltzman (1962). The equations are:

$$\dot{x} = \sigma(y - x), \tag{5.32}$$

$$\dot{y} = x(\rho - z) - y, \tag{5.33}$$

$$\dot{z} = xy - \beta z, \tag{5.34}$$

where $\beta = 8/3$, $\sigma = 10$, and $\rho = 28$. The parameter σ can be interpreted as the Prandtl number, which is the ratio of viscous diffusion rate and thermal diffusion rate. The parameter ρ can be interpreted as the Rayleigh number determining the relative strengths of convection to conduction in heat transfer in the flow.

Determine the fixed points (x_0, y_0, z_0) of the Lorenz model.

Consider ρ as a free parameter. In linear stability analysis of the fixed points define $(x, y, z) = (x_0 + x', y_0 + y', z_0 + z')$. Solve the equations, retaining only linear terms in the small quantities (x', y', z') (hint: the solutions are of the form $(x', y', z') = (X, Y, Z) \exp(\lambda t)$), and determine for which values of ρ all the fixed points are unstable ($\lambda > 0$).

From the linear stability analysis above, using (3.11) and (3.12) show that any volume in the phase space of the Lorenz model will shrink to zero. The phase space trajectories will approach a strange attractor, the Lorenz attractor, with a fractal dimension smaller than 3.

* From numerical simulation determine the fractal dimension of the Lorenz attractor by the Grassberger–Procaccia method (Grassberger & Procaccia, 1983).

* From numerical simulation, measure the mean exponential separation of nearby trajectories, see Figures 5.2 and 5.3, and calculate the Lyapunov exponent for the Lorenz model.

6

Helicity

Few exact results regarding fully developed turbulence have yet been derived, the most celebrated being Kolmogorov's four-fifth law. The four-fifth law is based on the fact that energy is transferred through the inertial range from the integral scale to the dissipation scale. The four-fifth law $\langle \delta u(l)_{\parallel}^3 \rangle = -(4/5)\bar{\varepsilon}\, l$ states that the third order correlator associated with energy flux equals the mean energy dissipation. The derivation of the four-fifth law relies on the energy being an inviscid invariant of the flow. In a 3D flow there exist other inviscid invariants, one being the helicity. The helicity density is defined as $h = u_i \omega_i$, and the helicity is $H = \int h\, \mathrm{d}\mathbf{x}$. As the name indicates, the helicity is a measure of the density of helices in the flow. Since right handed helices and left handed helices are measured with opposite signs, the helicity is a measure of the extent of mirror asymmetry in a flow. Typically the mirror symmetry is broken in flows for which a rotation, like the spinning of the Earth, introduces a preferred direction of spinning of the vortices in the flow.

6.1 The helicity spectrum

It follows from the time derivative, (A.35) and (A.36),

$$\frac{\mathrm{d}H}{\mathrm{d}t} = -2\nu \int \partial_j \omega_i \partial_j u_i \mathrm{d}\mathbf{x}$$

$$= -2\nu (2\pi)^3 \, \iota\, \epsilon_{ijl} \int k^2 k_j u_l(\mathbf{k}) u_i^*(\mathbf{k}) \mathrm{d}\mathbf{k}, \tag{6.1}$$

that the helicity H is a quadratic inviscid invariant of the NSE.

Due to the conservation of helicity for $\nu = 0$, a relation similar to the four-fifth law exists for the transfer of helicity (Chkhetiani, 1996; L'vov et al., 1997). This leads to another scaling relation for a third order correlator associated with the flux of helicity, $\langle \delta \mathbf{u}_{\parallel}(\mathbf{l}) \cdot [\mathbf{u}_{tt}(\mathbf{r}) \times \mathbf{u}_{tt}(\mathbf{r}+\mathbf{l})] \rangle = (2/15)\bar{\delta}\, l^2$, where $\bar{\delta}$ is the mean dissipation of helicity. The tensor index l (longitudinal) refers to the direction along $\mathbf{l}$, while

95

index t (transversal) refers to a direction perpendicular to **l**. In isotropic flow the correlator depends only on the distance $l = |\mathbf{l}|$ between the points. This relation is called the *two-fifteenth law*, due to the numerical factor. It establishes another nontrivial scaling relation for velocity differences in a turbulent helical flow. The two-fifteenth law is derived in the same way as the four-fifth law, and the interested reader is referred to L'vov *et al.* (1997).

In helical turbulence, coexisting cascades of energy and helicity were envisaged by Brissaud *et al.* (1973). Based on dimensional analysis it was conjectured that the helicity cascade is "linear" in the sense that the spectral helicity density is proportional to the spectral energy density: $H(k) \propto E(k) \propto k^{-5/3}$. This scenario was supported numerically by André and Lesieur (1977) in an Eddy-Damped Quasi-Normal Markovian (EDQNM) closure calculation and by Borue and Orszag (1997) in a direct numerical simulation. Following Brissaud *et al.* (1973), the existence of a linear helicity cascade is due to the fact that it is the same distortion time for eddies at a given scale that leads to nonlinear transfer of both energy and helicity. The time for distorting a vortex of spatial scale k^{-1} depends on the density of vortices at larger scales, which is represented by the enstrophy spectrum $Z(p) = p^2 E(p)$ at spectral scales $0 < p < k$. The distortion time at a scale k is then estimated as (Kraichnan, 1971)

$$\tau_k \sim \left(\int_0^k p^2 E(p) \mathrm{d}p \right)^{-1/2}. \tag{6.2}$$

Here, as before, $\sim$ denotes "equal within order unity factors." The distortion time τ_k signifies the time over which the energy at scale k is transferred, so for the energy we have

$$\Pi_E(k) \sim k E(k)/\tau(k), \tag{6.3}$$

and similarly for helicity

$$\Pi_H(k) \sim k H(k)/\tau(k). \tag{6.4}$$

Assuming a scaling relation $E(k) \sim \bar{\varepsilon}^x k^y$, (6.2) and (6.3) can be solved for x, y and the K41 result

$$E(k) \sim \bar{\varepsilon}^{2/3} k^{-5/3} \tag{6.5}$$

follows, where $\bar{\varepsilon}$ is the mean energy dissipation, or mean nonlinear energy transfer, or mean energy input. The restriction $y > -3$ ensures that the integral in (6.2) is convergent in the lower limit. Correspondingly, from (6.2) and (6.4) we obtain

$$H(k) \sim \bar{\delta} \bar{\varepsilon}^{-1/3} k^{-5/3}, \tag{6.6}$$

where $\overline{\delta}$ is the mean helicity input. The linear helicity cascade is derived under the assumption that helicity dissipation is negligible in the inertial range.

6.2 Comparison with 2D turbulence

Before considering the dual cascades of energy and helicity (Ditlevsen & Giuliani, 2001a), we briefly return to the case of 2D turbulence. In two dimensions, helicity is identically zero because the vorticity is always perpendicular to the velocity field. In 2D flow, the integral of the squared vorticity, the enstrophy, is also a quadratic inviscid invariant; see Section 2.1. Here the coexistence of cascades of energy and helicity is prohibited in the limit $\text{Re} \to \infty$. The reason for this is that the enstrophy dominates at small scales, so that the ratio of energy dissipation to enstrophy dissipation vanishes for high Reynolds number flow. The inner scale k_Z^{-1} for enstrophy dissipation is determined from the energy spectrum $E(k) \sim k^{-3}$ and the kinematic viscosity ν by

$$\overline{\zeta} = \nu \int_0^{k_Z} k^4 E(k) \, \mathrm{d}k \sim \nu k_Z^2,$$

where $\overline{\zeta}$ is the mean dissipation of enstrophy. From this it follows that $k_Z \sim \nu^{-1/2}$. The energy dissipation is $\overline{\varepsilon} = \nu \int_0^{k_Z} k^2 E(k) \, \mathrm{d}k \sim \nu \log k_Z \sim -(1/2)\nu \log \nu \to 0$ as $\nu \to 0$. Consequently, energy is cascaded upscale in 2D turbulence.

The existence of simultaneous cascades of energy and helicity in 3D turbulence is a little surprising because the same type of dimensional argument as for the cascades of energy and enstrophy in 2D turbulence naïvely applies. The apparent paradox is related to the fact that helicity is not a positive quantity, which can lead to very different scaling behaviors for even and odd powers of the velocity field (Biferale *et al.*, 1998; Ditlevsen & Giuliani, 2000). With respect to the scaling of the helicity spectrum, we therefore have to be very careful when applying dimensional arguments. This is most strongly manifested in the fact that the four-fifth law and the two-fifteenth law have completely different scaling for two correlators which have the same dimensionality but different tensorial structure.

6.3 The helicity dissipation scale

The helicity density can be represented spectrally by expanding the velocity vector $u_i(\mathbf{k})$ on the basis of *helical modes* (Waleffe, 1992). The helical modes $\mathbf{h}_\pm$ are the (complex) eigenvectors of the curl operator (A.16):

$$\iota \mathbf{k} \times \mathbf{h}_\pm = \pm k \, \mathbf{h}_\pm. \tag{6.7}$$

The vectors $\hat{\mathbf{k}} = \mathbf{k}/k, \mathbf{h}_+, \mathbf{h}_-$ form an orthonormal basis, $\hat{\mathbf{k}} \cdot \mathbf{h}_\pm = \mathbf{h}_+ \cdot \mathbf{h}_-^* = 0$, and $\hat{\mathbf{k}} \cdot \hat{\mathbf{k}} = \mathbf{h}_+ \cdot \mathbf{h}_+^* = \mathbf{h}_- \cdot \mathbf{h}_-^* = 1$. It is easy to see that in coordinates with $\mathbf{k}$ along the first axis they read

$$[\hat{\mathbf{k}}, \mathbf{h}_+, \mathbf{h}_-] = \left[\left(1,0,0\right), \left(0, 1/\sqrt{2}, \iota/\sqrt{2}\right), \left(0, \iota/\sqrt{2}, 1/\sqrt{2}\right) \right]. \qquad (6.8)$$

Using incompressibility, $\mathbf{k} \cdot \mathbf{u}(\mathbf{k}) = 0$, we have $\mathbf{u}(\mathbf{k}) = u_+(\mathbf{k})\mathbf{h}_+ + u_-(\mathbf{k})\mathbf{h}_-$ and the energy and helicity in the mode $\mathbf{u}(\mathbf{k})$ are

$$E(\mathbf{k}) = \mathbf{u}(\mathbf{k}) \cdot \mathbf{u}(\mathbf{k})^*/2 = (|u_+(\mathbf{k})|^2 + |u_-(\mathbf{k})|^2)/2, \qquad (6.9)$$

and

$$H(\mathbf{k}) = \mathbf{u}(\mathbf{k}) \cdot \omega(\mathbf{k})^*/2 = k(|u_+(\mathbf{k})|^2 - |u_-(\mathbf{k})|^2)/2. \qquad (6.10)$$

Thus the spectral energy and helicity densities can be separated into the densities of positive and negative helicity:

$$E(k) = E_+(k) + E_-(k), \qquad (6.11)$$

$$H(k) = H_+(k) + H_-(k) = k[E_+(k) - E_-(k)]. \qquad (6.12)$$

From this we have the rigorous constraint on the spectral helicity density

$$|H(k)| \leq kE(k). \qquad (6.13)$$

A similar constraint can be derived regarding the mean inputs of energy $\bar{\varepsilon}$ and helicity $\bar{\delta}$. Suppose the flow is subject to a force $\mathbf{f}$ with $\mathbf{f}(\mathbf{k}) = 0$ for $|\mathbf{k}| > K$, where K is a wave number at the integral scale. Then it follows that $|\bar{\delta}| \leq K\bar{\varepsilon}$ (Borue & Orszag, 1997). Solving (6.11) and (6.12) with respect to $E_+(k)$ and $E_-(k)$, and using the scaling relations (6.5) and (6.6), gives

$$E_+(k) = \left(C\bar{\varepsilon}^{2/3} k^{-5/3} + C_H \bar{\delta}\bar{\varepsilon}^{-1/3} k^{-8/3} \right)/2, \qquad (6.14)$$

$$E_-(k) = \left(C\bar{\varepsilon}^{2/3} k^{-5/3} - C_H \bar{\delta}\bar{\varepsilon}^{-1/3} k^{-8/3} \right)/2, \qquad (6.15)$$

where C and C_H are some (non-universal) Kolmogorov constants of order one.

The energy dissipation is given as

$$D_E = v \int_0^\infty k^2 E(k)\, \mathrm{d}k \approx v \int_0^{k_E} k^2 E(k)\, \mathrm{d}k, \qquad (6.16)$$

where k_E is the (inverse) Kolmogorov scale. For $k > k_E$ the viscosity makes the energy density $E(k)$ exponentially suppressed. The Kolmogorov scale is (1.11)

$$k_E = 1/\eta \sim (\bar{\varepsilon}/v^3)^{1/4}. \qquad (6.17)$$

The dissipation is linear in $E(k)$ and thus it can be split into dissipation of the positive and negative helicity parts of the spectrum. This implies that the dissipation of one sign of helicity $(s = \pm)$ is

$$D_{H_s} = \nu \int_0^\infty k^2 H_s(k) \, dk \sim \nu \int_0^{k_H} k^2 H_s(k) \, dk$$

$$= \nu \int_0^{k_H} k^3 E_s(k) \, dk \sim \nu k_H^4 E_s(k_H); \qquad (6.18)$$

see (A.37). Using the fact that $E_s(k)$ is scaling according to K41 theory, the helicity of sign s is dissipated at a scale determined by

$$D_{H_s} \sim \nu k_H^4 E_s(k_H) \sim \nu k_H^4 \left(\bar{\varepsilon}^{2/3} k_H^{-5/3} + s \bar{\delta} \bar{\varepsilon}^{-1/3} k_H^{-8/3} \right) \sim \bar{\delta}. \qquad (6.19)$$

The last term on the right hand side of (6.19) can be estimated by noting that the mean energy and helicity inputs are identical to the mean energy and helicity dissipations, thus if κ is a wave number at the integral scale where the energy input at wave numbers larger than κ is negligible, we have $\bar{\delta} \le \kappa \bar{\varepsilon}$ and the second term is less than κ/k_H times the first term. Neglecting the second term, we get an (inverse) inner scale, k_H, for dissipation of helicity

$$k_H \sim \left[\bar{\delta}^3 / (\nu^3 \bar{\varepsilon}^2) \right]^{1/7}. \qquad (6.20)$$

This inner scale is different from the Kolmogorov scale k_E. The scale cannot be obtained by pure dimensional counting in a manner similar to the way that the Kolmogorov scale is obtained: $k_E \sim \bar{\varepsilon}^\alpha \nu^\beta$ implies that $(\alpha, \beta) = (1/4, 3/4)$. In the case of helical turbulence we can define an (integral) length scale $L = \bar{\varepsilon}/\bar{\delta}$, and thereby $k_H \sim \bar{\varepsilon}^\alpha \nu^\beta (k_H \bar{\varepsilon}/\bar{\delta})^\gamma$, from which γ is under-determined by dimensional counting. For $\gamma = 0$, the Kolmogorov scale is obtained, and for $\gamma = -3/4$, Equation (6.20) is obtained.

For any flow realization we must have $k_H \le k_E$, so a pure helicity cascade is not possible. This result can be obtained by estimating where the flow should be forced in order to dissipate the helicity at the Kolmogorov scale, so that $k_H \sim k_E$. Pumping helicity into the flow at wave number κ implies $\bar{\delta} \sim \kappa \bar{\varepsilon}$, and assuming $k_H \sim k_E$ we have

$$\left(\frac{(\kappa \bar{\varepsilon})^3}{\nu^3 \bar{\varepsilon}^2} \right)^{1/7} \sim \left(\frac{\bar{\varepsilon}}{\nu^3} \right)^{1/4}, \qquad (6.21)$$

which implies that $\kappa \sim k_E$. This shows that the flow must be forced at the Kolmogorov scale, which is in conflict with the existence of an inertial range. A similar result was obtained by Olla (1998) in a different way, using an argument based on the EDQNM approximation.

Furthermore, we have $k_H/k_E \propto v^{-3/7+3/4} = v^{9/28} \to 0$ as $v \to 0$. So again, for high Reynolds number helical flow the small scales will always be non-helical. The inner scale for helicity dissipation plays a different role in helical turbulence than the Kolmogorov scale. The dissipation of one sign of helicity at a given wave number will grow with wave number as $D_{H_s}(k) \propto k^{7/3}$, thus the dissipation of either sign of helicity will grow with wave number in the range $k_H < k < k_E$. This is only possible if there is a production of equal amounts of positive and negative helicity and a detailed balance between dissipation of positive and negative helicity in that range.

The scenario this analysis proposes for high Reynolds number helical turbulence then becomes the following. At the integral scale K, energy and helicity are forced into the flow. In the inertial range $K < k < k_H$ there is a coexisting cascade of energy and helicity where helicity follows a "linear cascade," with an $H(k) \sim k^{-5/3}$ spectrum. In the range $k_H < k < k_E$ the dissipation of helicity dominates, with a detailed balance between dissipation of positive and negative helicity, and the right–left symmetry of the flow is restored. The balanced positive and negative helicity are generated in analogy to the enstrophy being generated in high Reynolds number flow. The proposed scenario has been illustrated in a shell model of turbulence (Ditlevsen & Giuliani, 2001b). However, since the considerations presented here are purely phenomenological they should be tested in experiments or numerical simulations of the NSE.

6.4 Helicity in shell models

In order to test these ideas in a model system, one can investigate the role of helicity and the structure of the helicity transfer in a shell model (Ditlevsen, 1997).

Shell models lack any spatial structures, so that only certain aspects of turbulent cascades have meaningful analogies in shell models. This should be kept in mind, especially when studying helicity, which is intimately linked to spatial structures. So the following, concerns only the spectral aspects of the helicity, and energy cascades. In the following the GOY model with the standard parameters $(\epsilon, \lambda) = (1/2, 2)$ is used.

The energy flux is defined in the usual way as

$$\Pi_n^{(E)} = -d/dt|_{\text{n.l.}} \sum_{m=1}^{n} E_m, \qquad (6.22)$$

where $d/dt|_{\text{n.l.}}$ is the time rate of change due to the nonlinear term in (3.22). The helicity flux $\Pi_n^{(H)}$ is defined similarly, and we have the expressions (3.45) and (3.46) for the fluxes

$$\langle \Pi_n^{(E)} \rangle = \Delta_{n+1} + \Delta_n/2 = \bar{\varepsilon}, \tag{6.23}$$

$$\langle \Pi_n^{(H)} \rangle = (-1)^n k_n(\Delta_{n+1} - \Delta_n) = \bar{\delta}, \tag{6.24}$$

in the inertial range of n, where $\Delta_n = k_{n-1}\Im\langle u_{n-1}u_n u_{n+1} \rangle$, $k_n = \lambda^n$, and $\bar{\varepsilon}, \bar{\delta}$ are the mean dissipations of energy and helicity, respectively. The left equalities hold without averaging as well. These equations are the shell model equivalents of the four-fifth and the two-fifteenth laws.

In the shell model we have the mean energy input

$$\bar{\varepsilon} = \sum_n \Re\langle f_n u_n^* \rangle, \tag{6.25}$$

and mean helicity input

$$\bar{\delta} = 2\sum_n (-1)^n k_n \Re\langle f_n u_n^* \rangle, \tag{6.26}$$

where f_n is the force at shell number n. In complete analogy with Navier–Stokes turbulence we have $|\bar{\delta}| \le 2k_f\bar{\varepsilon}$, where k_f is the largest wave number such that $f_k = 0$ for $k > k_f$, provided that there is a non-negative mean energy injection for all wave numbers $k \le k_f$. The force can be chosen in many ways. A simplifying choice is $f_n = f_n^{(0)}/u_n^*$, with $f_n^{(0)}$ independent of the shell velocities. Then we have $\bar{\varepsilon} = \sum_{n<n_f} f_n^{(0)}$ and $\bar{\delta} = \sum_{n<n_f}(-1)^n k_n f_n^{(0)}$, where n_f indicates the end of the integral scale. By choosing the coefficients, which may be stochastic or deterministic functions of time, this last sum can vanish identically, which is referred to as helicity free forcing. The third order correlation function $S_n^3 \equiv -\Im\langle u_{n-1}u_n u_{n+1} \rangle = -\Delta_n/k_{n-1}$ is obtained from (6.23) and (6.24):

$$k_n S_n^3 = \frac{4}{3}\left(\bar{\varepsilon} - (-1)^n \bar{\delta}/k_n\right). \tag{6.27}$$

The last term in the parenthesis is sub-leading with period two oscillations. When $\bar{\delta} = 0$ the sub-leading term disappears and the scaling from the equivalent of the four-fifth law (6.23) is obtained, see Figure 6.1. The simulations are performed with the force $f_2^{(0)} = 10^{-2}(1 + \iota)$ and $f_3^{(0)} = -Af_2^0/\lambda$ with $A = 1$ and $A = 0$, corresponding to $(\bar{\varepsilon}, \bar{\delta}) = (0.01, 0)$ (diamonds) and $(\bar{\varepsilon}, \bar{\delta}) = (0.01, 0.08)$ (crosses), respectively.

Helicity is dissipated with opposite signs for odd and even shells. If we consider the third order structure function associated with the helicity transfer as defined by (6.24), we see period two oscillations growing with n (Figure 6.2). This period two oscillation is due to the dissipation.

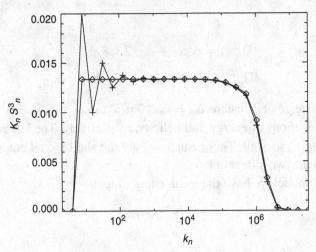

Figure 6.1 The third order structure function S_n^3 in the cases $\bar{\delta} > 0$ (crosses) and $\bar{\delta} = 0$ (diamonds). In the case of helicity free forcing the period 2 oscillations disappear. In the two runs we have 25 shells, $\nu = 10^{-9}$, $f_n = 0.01(1+\iota)(\delta_{n,2}/u_2^* - A\delta_{n,3}/2u_3^*)$ with $A = 0, 1$, respectively.

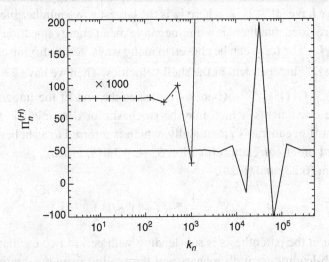

Figure 6.2 The helicity flux $\langle \Pi_n^{(H)} \rangle$ in the case $\bar{\delta} > 0$. The same curve is multiplied by 1000 and over-plotted in order to see the inertial range. The period 2 oscillations in the helicity transfer come from the helicity dissipation.

The helicity flux is given by (6.24) for the range of scales $K < k < k_H$, where the helicity dissipation is negligible. For scales $k_H < k < k_E$ we have

$$\langle \Pi_n^{(H)} \rangle = \bar{\delta} - \langle D_n^{(H)} \rangle, \tag{6.28}$$

where $D_n^{(H)}$ is the helicity dissipation at shells $m \leq n$:

$$D_n^{(H)} = 2v \sum_{m=1}^{n} (-1)^m k_m^3 |u_m|^2. \tag{6.29}$$

In the inertial range for energy transfer we have the Kolmogorov scaling $u_n \sim k_n^{-1/3}$, so the helicity dissipation can be estimated:

$$D_n^{(H)} \sim 2v \sum_{m=1}^{n} (-1)^m k_m^{7/3} \sim \lambda^{7/3} \frac{(-1)^n \lambda^{7n/3} - 1}{\lambda^{7/3} + 1} \sim (-1)^n k_n^{7/3}. \tag{6.30}$$

This is the shell model equivalent of (6.19) if n is at the Kolmogorov scale. Figure 6.3 shows $|\langle \Pi_n^{(H)} \rangle|$ and $\langle \Pi_n^{(E)} \rangle$ as functions of wave number. The scaling (6.30) of the helicity dissipation is the straight line, the horizontal dashed line is $\bar{\delta}$. The inertial range for helicity transfer is to the left of the crossing of the two lines. The crossing is the inner scale for helicity transfer k_H, which does not coincide with the Kolmogorov scale k_E.

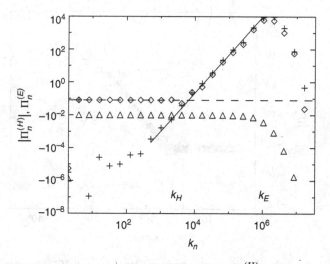

Figure 6.3 The absolute values of the helicity flux $|\langle \Pi_n^{(H)} \rangle|$ (diamonds) show a crossover from the inertial range for helicity to the range where the helicity is dissipated. The line has a slope of 7/3, indicating the helicity dissipation. The dashed line indicates the helicity input $\bar{\delta}$. The crosses are the helicity flux in the case $\bar{\delta} = 0$ where there is no inertial range and k_H coincides with the integral scale. The triangles are the energy flux $\langle \Pi_n^{(E)} \rangle$.

The "pile-up" for k larger than k_H was earlier interpreted as a bottleneck effect (Biferale *et al.*, 1998). It is rather a balance between positive and negative helicity dissipation, and does not contribute to the dissipation of the injected helicity.

The force $f_n = f_n^{(0)}/u_n^*$ can potentially cause numerical trouble when $|u_n|$ becomes small. It is easy to see that the linear equation for (real) shell velocity u_n, neglecting the nonlinear transfer, $\dot{u}_n = f/u_n$, will create a finite time singularity. This is not the case for the force suggested by Olla (1998) at two shells, $f_n = (aE_{n+1}u_n)/(E_n + E_{n+1})$ and $f_{n+1} = (bE_nu_{n+1})/(E_n + E_{n+1})$, where a and b are constants determining the ratio of energy to helicity input. The coupled set of equations $\dot{u}_n = f_n$, $\dot{u}_{n+1} = f_{n+1}$, is integrable by solving for $y = u_n/u_{n+1}$, and it has no finite time singularities.

In order to verify the scaling relation between k_H and $\bar{\delta}$, a set of simulations with constant energy input $\bar{\varepsilon} = 0.01$ and varying helicity input $\bar{\delta} = (0.0001, 0.001, 0.005, 0.01, 0.08)$ has been performed. In Figure 6.4 the spectra of the absolute value of the helicity transfer normalized with $\bar{\delta}$ are plotted versus wave number normalized with k_H. In each case k_H is calculated from (6.20), and a clear data coincidence is seen.

Thus we can conclude that there is an inner scale for helicity dissipation in the GOY shell model. This inner scale is always larger than the Kolmogorov scale.

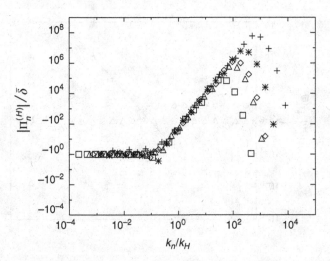

Figure 6.4 Five simulations with constant viscosity $\nu = 10^{-9}$, constant energy input $\bar{\varepsilon} = 0.01$, and varying helicity input $\bar{\delta} = (0.0001, 0.001, 0.005, 0.01, 0.08)$ are shown. The absolute values of the helicity flux $|\langle \Pi_n^{(H)} \rangle|$ divided by $\bar{\delta}$ are plotted against the wave number divided by $k_H = (\nu^3 \bar{\varepsilon}^2 / \bar{\delta}^3)^{-1/7}$, which is obtained from (6.20) neglecting $O(1)$ constants. A clear data collapse is seen.

Thus there may exist two inertial ranges in helical turbulence: a range smaller than k_H with coexisting cascades of energy and helicity where both the four-fifth and the two-fifteenth law apply; and a range between k_H and k_E where the flow is non-helical and only the four-fifth law applies. These findings should be investigated in observations of helical flow or in direct numerical simulations of the NSE with helical forcing.

6.5 A generalized helical GOY model

The way helicity appears in the shell model with opposite signs for every other shell has no direct correspondence with a real flow. In order to make a more realistic model for helicity, Kerr and Biferale (1995) constructed a model consisting of two coupled replicas of the GOY model. They noted that within the shell model each shell is maximally helical, that is $|H_n| = k_n E_n$, implying that the velocity and the vorticity are parallel. For the NSE this would correspond to each Fourier mode being proportional to either of the helical modes (6.8) h_+ or h_-. In order to avoid this restriction, each shell velocity must contain two components

$$u_n = u_n^+ h_n^+ + u_n^- h_n^-, \tag{6.31}$$

where $h_n^\pm$ are the spectral representations at shell n of the helical modes and $u_n^\pm$ are the two shell velocities. The governing equation should be specified for the two shell velocities, which is done in the usual way by demanding only nearest neighbor triads and incompressibility in phase space (Liouville's theorem). This can be done in four different ways, as shown in Figure 6.5.

In the same way as for the GOY model, the conservation laws are obtained by "telescoping," such that only one type of triad interaction can be involved in a conservation law. The construction of the model is straightforward, with σ_i denoting the sign, we get two equations similar to (3.22):

$$\dot{u}_n^+ = \iota k_n \left(u_{n+1}^{\sigma_1} u_{n+2}^{\sigma_2} - \frac{\epsilon}{\lambda} u_{n-1}^{\sigma_3} u_{n+1}^{\sigma_4} + \frac{\epsilon - 1}{\lambda^2} u_{n-2}^{\sigma_5} u_{n-1}^{\sigma_6} \right)^*$$
$$- \nu k_n^2 u_n^+ + f_n^+,$$

$$\dot{u}_n^- = \iota k_n \left(u_{n+1}^{-\sigma_1} u_{n+2}^{-\sigma_2} - \frac{\epsilon}{\lambda} u_{n-1}^{-\sigma_3} u_{n+1}^{-\sigma_4} + \frac{\epsilon - 1}{\lambda^2} u_{n-2}^{-\sigma_5} u_{n-1}^{-\sigma_6} \right)^*$$
$$- \nu k_n^2 u_n^- + f_n^-. \tag{6.32}$$

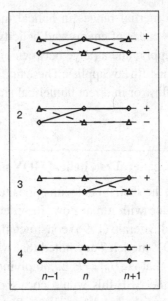

Figure 6.5 The four possible helical shell models.

The four models are defined by the set $(\sigma_1,\ldots,\sigma_6) = (-,+,-,-,+,-), (-,-,-,+,-,+), (+,-,+,-,-,+), (+,+,+,+,+,+)$, respectively. In this form the two conserved quantities

$$E_{1,2} = \sum_n k_n^{\alpha_{1,2}} \left(|u_n^+|^2 + |u_n^-|^2 \right), \tag{6.33}$$

with $\lambda^{\alpha_1} = 1$ and $\lambda^{\alpha_2} = 1/(\epsilon - 1)$, are conserved for any ϵ in all four models. We define a generalized helicity as

$$H = \sum_n k_n^{\alpha} \left(|u_n^+|^2 - |u_n^-|^2 \right). \tag{6.34}$$

The generalized helicity conservation leads to four different equations for the four models. It is easy to see that by taking the time derivative of H we get an equation with six terms for each shell. These must cancel in pairs of three, after having identified the three terms proportional to $u_{n-1}^{\sigma_1} u_n^{\sigma_2} u_{n+1}^{\sigma_3}$ in the sum over shells involved in a triad interaction. For the four models this give the equations:

$$\text{model 1: } 1 + \lambda^{\alpha}\epsilon + (\lambda^{\alpha})^2(\epsilon - 1) = 0,$$

$$\text{model 2: } 1 + \lambda^{\alpha}\epsilon - (\lambda^{\alpha})^2(\epsilon - 1) = 0,$$

$$\text{model 3: } 1 - \lambda^{\alpha}\epsilon - (\lambda^{\alpha})^2(\epsilon - 1) = 0,$$

$$\text{model 4: } 1 - \lambda^{\alpha}\epsilon + (\lambda^{\alpha})^2(\epsilon - 1) = 0. \tag{6.35}$$

The two solutions for λ^α in the four models are:

$$\text{model 1: } (-1, -1/(1-\epsilon)),$$

$$\text{model 2: } (\epsilon \pm \sqrt{\epsilon^2 + 4(\epsilon - 1)})/2(\epsilon - 1),$$

$$\text{model 3: } -(\epsilon \pm \sqrt{\epsilon^2 + 4(\epsilon - 1)})/2(\epsilon - 1)),$$

$$\text{model 4: } (1, 1/(1-\epsilon)). \tag{6.36}$$

Since any linear combination of the conserved quantities is also conserved, we see that model 4 and model 1 conserve energy and helicity separately for two sets of shells, thus they are equivalent and just a set of two independent GOY models. This result is obvious by looking at Figure 6.5, and substituting $u_{2n+1}^+ \leftrightarrow u_{2n+1}^-$. The solutions for the nontrivial models 2 and 3 are shown as functions of ϵ in Figure 6.6.

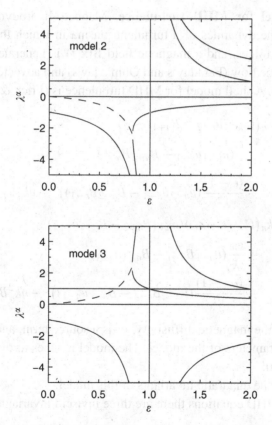

Figure 6.6 The set of values for λ^α as a function of ϵ for the helical shell models 2 and 3. The dashed lines show the real part of λ^α where it is complex.

The models resembling 3D turbulence will have energy as the only positive conserved quantity and three helicity-like quantities. The models resembling 2D turbulence will conserve two positive quantities, energy, and enstrophy and two additional helicity-like quantities.

6.6 Problems

6.1 Consider a forced system of two shells governed by the equations

$$\dot{u}_1 = (aE_2 u_1)/(E_1 + E_2),$$

$$\dot{u}_2 = (bE_1 u_2)/(E_1 + E_2),$$

where $E_i = |u_i|^2$ and a and b are constants determining the ratio of energy to helicity input. Show that this coupled set of equations, is integrable and has no finite time singularities. Hint: solve for $y = u_1/u_2$.

6.2 Shell model for MHD turbulence. Magneto-hydrodynamic turbulence describes the dynamics of a turbulent plasma in which the fluid is subject to Lorentz forces, and a magnetic field $\mathbf{B}(\mathbf{x}, t)$ is generated by the electric current of the flow (Faraday's and Ohm's laws) and advected by the velocity field $\mathbf{u}(\mathbf{x}, t)$. A shell model for MHD turbulence is (Frick & Sokoloff, 1998)

$$\dot{u}_n = \iota k_n \big((u_{n+1} u_{n+2} - B_{n+1} B_{n+2})$$

$$- \frac{\epsilon}{\lambda} (u_{n-1} u_{n+1} - B_{n-1} B_{n+1})$$

$$+ \frac{(\epsilon - 1)}{\lambda^2} (u_{n-2} u_{n-1} - B_{n-2} B_{n-1}) \big)^* - \nu k_n^2 u_n + f_n, \qquad (6.37)$$

$$\dot{B}_n = \iota k_n \big((1 - \epsilon - \epsilon_m)(u_{n+1} B_{n+2} - B_{n+1} u_{n+2})$$

$$+ \frac{\epsilon_m}{\lambda} (u_{n-1} B_{n+1} - B_{n-1} u_{n+1})$$

$$- \frac{(\epsilon_m - 1)}{\lambda^2} (u_{n-2} B_{n-1} - B_{n-2} u_{n-1}) \big)^* - \eta k_n^2 B_n + g_n, \qquad (6.38)$$

where η is the magnetic diffusivity, g_n is a source term, and λ, ϵ, and ϵ_m are the free parameters of the model. The model reduces to the GOY model for $B_n = g_n = 0$.

Determine the quadratic invariants of this model.

* For the MHD equations there are three inviscid invariants, total energy:

$$E = \int (\mathbf{u}^2 + \mathbf{B}^2) d\mathbf{x},$$

cross helicity:

$$H_C = \int \mathbf{u} \cdot \mathbf{B} dx,$$

and magnetic helicity:

$$H_B = \int \mathbf{A} \cdot \mathbf{B} dx,$$

where $\mathbf{B} = \nabla \times \mathbf{A}$. Compare these with the results found for the shell model above. Determine for $\lambda = 2$ the values for the parameters ϵ and ϵ_m which give the correct physical dimensions for the inviscid invariants (in 3D).

7

Intermittency

Intermittency in turbulence is a topic which has been actively investigated for several decades, and a major part of Frisch's book (1995) is devoted to the subject.

Dynamical systems are often characterized by long quiescent periods interrupted by bursts of activity. This kind of dynamics is called intermittent. A way of quantifying this could be by high pass filtering the dynamical signal. If the signal has purely Gaussian statistics, which would be natural for a system of many degrees of freedom, high pass filtering is a linear operation and the high pass filtered signal would be Gaussian as well. If the high pass signal differs from the Gaussian by having heavier tails, it is intermittent. Thus intermittency could be formally defined by the deviation from Gaussian statistics. In this case, intermittency could be a sign of dynamics not merely governed by simple statistics given by the central limit theorem or equilibrium statistical mechanics. In the case of a turbulent velocity field, high pass filtering roughly corresponds to extracting information on velocity differences below or at some cutoff length scale. As described previously, the statistics of velocity increments in turbulence is found not to be Gaussian. The self-similarity of the flow assumed in K41 theory is not valid and there will be corrections to the scaling exponents for the moments of the velocity increments as expressed in (1.64).

7.1 Kolmogorov's lognormal correction

The dynamical origin of the deviation from the K41 value for the scaling exponents is a very non-uniform distribution of the energy dissipation. Consider the mean energy dissipation within a sphere located at $\mathbf{x}$ of radius l:

$$\varepsilon_l(\mathbf{x}) = \frac{\nu}{\frac{4}{3}\pi l^3} \int_{|\mathbf{y}-\mathbf{x}|<l} \frac{1}{2}(\partial_j u_i + \partial_i u_j)^2 \mathrm{d}\mathbf{y}. \tag{7.1}$$

110

If the general scaling ansatz is valid, the qth moment of the energy dissipation – coarse grained in the form of (7.1) – should follow a scaling relation

$$\langle \varepsilon_l^q \rangle \sim l^{\tau(q)}. \tag{7.2}$$

The scaling exponent $\tau(q)$ is related to the scaling exponent $\zeta(p)$ for the structure functions (1.65). Using the K41 relation (1.8) we have for the pth moment structure function:

$$\langle \delta u(l)^p \rangle \sim \langle (\varepsilon_l l)^{p/3} \rangle \sim l^{p/3 + \tau(p/3)}. \tag{7.3}$$

so we have $\zeta(p) = p/3 + \tau(p/3)$. This relation holds if the scaling proposition (7.2) is valid. We assume that the dissipation in a given sphere of radius $\lambda^n L$ ($\lambda < 1$) is the result of a cascade of random independent fractionations of the energy flux from the large scale L. The energy dissipation can be viewed as a random variable $\varepsilon_{\lambda^n L} = x_L x_{\lambda L} \cdot \ldots \cdot x_{\lambda^n L}$ modeled as a product of independent random variables where each random factor $x_{\lambda^k L}$ represents the fraction of energy present at the sphere of radius $\lambda^{k-1} L$ which is cascaded into the sphere of radius $\lambda^k L$. The whole sequence of spheres includes the sphere of radius $\lambda^n L$. From the central limit theorem, it is natural to assume a lognormal distribution for such a process, since we have $\log \varepsilon_{\lambda^n L} = \log x_L + \log x_{\lambda L} + \cdots + \log x_{\lambda^n L}$, and the sum of independent random variables will tend to be normally distributed for $n \to \infty$. Kolmogorov (1962) and Obukhov (1962) formulated this idea and proposed the lognormal distribution for the energy dissipation, ε_l.

From energy conservation, the mean of the distribution $\langle \varepsilon_l \rangle = \bar{\varepsilon}$ is independent of l. Kolmogorov and Obukhov assumed that the variance scales with a parameter μ as

$$\langle (\log \varepsilon_l)^2 \rangle - \langle \log \varepsilon_l \rangle^2 \equiv \sigma^2 = \mu \log(L/l). \tag{7.4}$$

With this assumption the scaling exponents can be calculated directly. Using the short notation $x = \log \varepsilon$ and the lognormal distribution, we obtain

$$\langle \varepsilon^p \rangle = \frac{1}{\sigma \sqrt{2\pi}} \int e^{px} e^{-(x-\bar{x})^2/2\sigma^2} dx = e^{\sigma^2 p^2/2 + \bar{x} p}, \tag{7.5}$$

where the last equality is obtained by "completion of the square." For $p = 1$ we have $\bar{\varepsilon} \sim \exp(-\sigma^2/2 - \bar{x})$, which must be independent of l. Thus we must have

$\bar{x} = -\sigma^2/2$ and $\langle \varepsilon^p \rangle \sim \exp[-\sigma^2(p^2 - p)/2]$. By inserting the expression for the variance we get

$$\langle \varepsilon^p \rangle \sim l^{\mu(p-p^2)/2}. \tag{7.6}$$

This result, obtained by Kolmogorov and Obukhov, is only asymptotically correct for increasing l because the variance assumption (7.4) fails except for large l. A probabilistically correct derivation given in Appendix A.9 shows that

$$\langle \varepsilon_l^p \rangle = \exp\left\{ p\bar{\varepsilon} + \frac{1}{2}\log\left[1 + \frac{R(l)^3}{l^3}V^2 \right](p^2 - p) \right\}, \tag{7.7}$$

where V is the coefficient of variation (the standard deviation divided by the mean) of the limit of $\varepsilon_l(\mathbf{x})$ as $l \to 0$, and where $R(l)$ is a function that relates to the correlation coefficient function structure of the velocity field. The essential behavior of $R(l)$ is that it approaches a constant correlation length scale as $l \to \infty$, while it is asymptotically equal to l as $l \to 0$. Thus it follows that $\langle \varepsilon_l^p \rangle \sim \mathcal{O}(1)$ as $l \to 0$, showing that (7.6) is completely wrong for small l.

The argument by Kolmogorov and Obukhov that $\varepsilon_l(\mathbf{x})$ should have a lognormal distribution appears reasonable, but it cannot hold mathematically exact for all l. For large l, the integral over the sphere of radius l of a homogeneous random field represents a summation that in general tends to be Gaussian (not necessarily, though). Without referring to the central limit theorem, another argument is that the sum of lognormal random variables does not have lognormal distribution. However, assuming that the random integrand has a lognormal distribution with mean $\bar{\varepsilon}$ and coefficient of variation V for any $\mathbf{x}$, then $\varepsilon_l(\mathbf{x})$ has a lognormal distribution asymptotically as $l \to 0$, and a Gaussian distribution asymptotically as $l \to \infty$, under suitable conditions on the correlation structure. Since the lognormal distribution approaches the Gaussian distribution asymptotically as its coefficient of variation decreases, the lognormal distribution family is a good approximation for the interpolation between the two extreme cases of $l = 0$ and $l = \infty$. In fact, simulations show that for any l the lognormal distribution is a better approximation to the simulated distribution than the Gaussian distribution.

However, due to the experience that turbulence is much more intermittent than the sample functions of a stationary lognormal random process, the lognormal model is mainly of historical interest. The turbulence phenomenon appears as if one type of process is governing in the calm periods between the bursts of more violent turbulence, and another process is governing within the bursts. The lognormal model may be relevant within the burst periods, but averaging over time intervals and/or space domains that include the calm periods should be over an inhomogeneous distribution of some mixture type. Mixed distribution processes are thus a potential modeling tool for describing intermittent turbulence (Grigoriu *et al.*, 2003).

7.2 The β-model

The physical picture of the turbulent cascade being more and more rarefied as the dissipative scale is approached leads to an obvious modification of the Richardson picture of the cascade. The β-model states that during the cascade from a scale l to a scale λl, the smaller eddies will occupy only a fraction $0 < \beta < 1$ of the volume occupied by the eddies at scale l. There is thus a rarefaction of the active flow as the scales decrease. The probability density of finding an eddy at a given scale $l = \lambda^n l_0$ is then simply proportional to the active volume:

$$p_l = \beta^n = \beta^{\log(l/l_0)/\log(\lambda)} = \left(\frac{l}{l_0}\right)^{\log\beta/\log\lambda}. \tag{7.8}$$

The exponent $\log\beta/\log\lambda \equiv 3 - D$ can be interpreted as the co-dimension of the fractal containing the active flow (Mandelbrot, 1977). The fractal dimension of the active flow is thus D. To see this, recall the definition of the fractal dimension. The fractal or Hausdorff dimension is defined from the scaling of the number of spheres of radius λl necessary to cover the fractal in comparison to the number of spheres of radius l covering the fractal

$$\frac{n(\lambda l)}{n(l)} \sim \lambda^{-D}. \tag{7.9}$$

The probability density $p_{\lambda l}$ is simply the fraction of spheres of radius λl necessary to cover the fractal out of the number of spheres necessary to cover the full space

$$p_{\lambda l} = \frac{n(\lambda l)/n(l)}{N(\lambda l)/N(l)} = \lambda^{3-D}, \tag{7.10}$$

in 3D space. Now the intermittency corrections to the scaling can be calculated. Firstly, we can obtain the spectral energy flux Π_l of the flow through the scale l:

$$\Pi_l \sim \frac{\delta u(l)^3}{l} p_l \sim \delta u(l)^3 l^{2-D}. \tag{7.11}$$

Energy conservation in the inertial range gives $\Pi_l = \bar\varepsilon$, and the scaling of the velocity field immediately follows:

$$\delta u_l \sim \bar\varepsilon^{1/3} l^{-(2-D)/3} = \bar\varepsilon^{1/3} l^{1/3-(3-D)/3}. \tag{7.12}$$

Thus there is a correction to the scaling of the velocity field depending on the dilution of the active volume. For K41 we assume the energy dissipation to be space filling with $D = 3$, and the dimensional scaling exponent $h = 1/3$ appears. The structure functions are obtained similarly:

$$S_p(l) = \langle \delta u(l)^p \rangle \sim \int \delta u(l)^p p_l dx \sim l^{p(1/3-(3-D)/3)+(3-D)}$$

$$= l^{p/3+(3-D)(1-p/3)} \sim l^{\zeta(p)}. \tag{7.13}$$

The scaling exponent $\zeta(p) = p/3 + (3 - D)(1 - p/3)$ is then linear in p for the β-model. For the sixth order structure function we have $\zeta(6) = 2 - (3-D)$, so that the intermittency correction is exactly the co-dimension of the active fractal. The β-model introduces one extra scaling parameter, represented by β or the fractal dimension D, which leads to the linear intermittency correction.

7.3 The multi-fractal model

Comparing the result from the β-model with the observed curve connecting the points in Figure 1.5 led Frisch and Parisi (1985) to a description with a whole range of scaling parameters. The Navier–Stokes equation is invariant under the scaling transformation, $t, \mathbf{x}, \mathbf{u}, \nu \to \lambda^{1-h}t, \lambda\mathbf{x}, \lambda^h\mathbf{u}, \lambda^{1+h}\nu$, for any value of h. By imposing global scale invariance, the K41 theory derives the unique value $h = 1/3$ from dimensional counting. However, if only local scale invariance is assumed, we may assume a whole continuum of scaling exponents h, each of which describes the scaling within a fractal subset S_h with dimension $D(h)$. Within this subset we have $\delta u(l) \sim l^h$ and the structure functions can be calculated as

$$S_p(l) = \langle \delta u(l)^p \rangle = \int p_l(h) \, \delta u(l)^p d\mu(h) \sim \int l^{ph+3-D(h)} d\mu(h), \tag{7.14}$$

where $\mu(h)$ is a measure of the subset S_h. This integral cannot be evaluated without knowledge of the measure $\mu(h)$. However, the integral can be estimated in the limit $l \to 0$ by the saddle-point approximation. Since the saddle-point approximation is used extensively in the multi-fractal models, it is worthwhile to review it briefly. Consider an integral of the form

$$I = \int l^{f(x)} dx = \int e^{\log l f(x)} dx, \tag{7.15}$$

where the function $f(x)$ varies faster than logarithmic with x. Assume furthermore that $f(x)$ has a single minimum at x_0, so that in the neighborhood of x_0 we can write

$f(x) \approx f(x_0) + f''(x_0)(x-x_0)^2/2$ where $f''(x_0) > 0$. We are interested in the limit $l \to 0$, so we assume $\log l < 0$. The integral can now be evaluated:

$$I = \int e^{\log l f(x)} dx \approx e^{\log l f(x_0)} \int e^{-|\log l| f''(x_0)(x-x_0)^2/2} dx$$

$$= l^{f(x_0)} \sqrt{\frac{2\pi}{|\log l| f''(x_0)}}. \tag{7.16}$$

Taking the logarithm and dividing by $\log l$ gives

$$\frac{\log I}{\log l} = f(x_0) + \frac{1}{2\log l}[\log(2\pi) - \log|\log l| - \log f''(x_0)] \to f(x_0) \tag{7.17}$$

as $l \to 0$. In this limit we can neglect the logarithmic correction and obtain the scaling relation $I \sim l^{f(x_0)}$. Comparing this to (7.14), we see that the measure $\mu(h)$ only matters for the logarithmic correction, and in the limit $l \to 0$ we get

$$S_p(l) \sim l^{\zeta(p)}, \tag{7.18}$$

with

$$\zeta(p) = \inf_h [ph + 3 - D(h)]. \tag{7.19}$$

This relation shows that the scaling exponent $\zeta(p)$ is the one dimensional Legendre transform of $D(h) - 3$.

7.4 Intermittency in shell models

Even though the shell models do not have spatial structure, they exhibit a spectrum of anomalous scaling exponents very similar to fluid turbulence. Thus there is a deviation from the K41 equivalent predictions for shell models. The origin of intermittency in shell models is the irregularities in the temporal signals. The energy flux through the shells becomes more and more intermittent as the Kolmogorov scale is reached. This temporal intermittency can be visualized using Howmöller diagrams, plotting the energy flux in a contour plot as a function of time and shell number (Ditlevsen, 1996). In a Howmöller diagram, time is on the y-axis while, in this case, the shell number is on the x-axis. This is shown in Figure 7.1, where the dark areas represent high values of the energy flux. Bursts of energy are traveling through the inertial range with an increasing "speed," whereby the bursts become more and more focused in time as the dissipative scale is approached. Furthermore, it is seen in the Howmöller diagram that the peaks in the energy transfer have a tendency to split into two as they travel through the inertial range. This is a manifestation of the dynamics described qualitatively in the β-model.

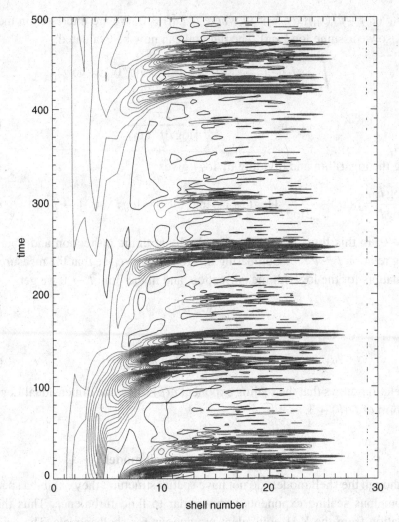

Figure 7.1 A Howmöller diagram showing energy transfer through the inertial range in the Sabra shell model. The contour lines are the energy transfer Π_n through shell n. The figure shows bursts of energy transfer with increasing "speed" toward the dissipation scale around $n = 25$. The energy dissipation is intermittent.

7.5 Probability densities and intermittency

The concentration of the energy flux through the inertial range is seen in the probability densities for the flux through shell n being more and more concentrated around zero with a more and more heavy (positive) tail as n increases. The skewness of the distributions is a necessary consequence of the forward energy cascade, since the energy flux has a positive mean. Figure 7.2 shows the central parts of the

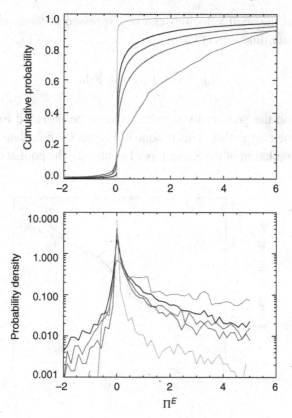

Figure 7.2 Results from a simulation of the standard 3D Sabra model with 30 shells and the parameters $f_n = (1 + i)\delta_{n,4}, \nu = 10^{-10}$. The top panel shows the cumulative probabilities and the lower panel shows the probability densities for the energy flux through the shells. The shells 5, 10, 15, 20, and 25 (light to dark) are plotted. Note that only the central part of the probabilities is plotted, the positive tail of the distribution will be heavier for the larger number shells.

distributions for shells 5, 10, 15, 20, and 25 (light to dark) obtained from a simulation of the standard 3D Sabra model with 30 shells. An important observation is that for the shell models the intermittency corrections to the scaling exponents are related to the dynamics in the inertial range and not merely a consequence of the way energy is dissipated.

The problem of determining the inertial range anomalous scaling exponents $\zeta(p)$ in a turbulent flow is similar to determining the scaling exponents in the shell models:

$$\langle |u_n|^p \rangle \sim k_n^{-\zeta(p)} = \lambda^{-n\zeta(p)}. \tag{7.20}$$

The moments of the shell velocities can be expressed in terms of the stationary probability density functions (PDFs):

$$\langle |u_n|^p \rangle = \int p_n(u) |u|^p \mathrm{d}u. \tag{7.21}$$

By this definition the probability densities should be obtained for the velocities rather than for the energy flux. This is done in Figure 7.3, calculated from the same (rather short) simulation of the Sabra model as above. The pooled distributions for

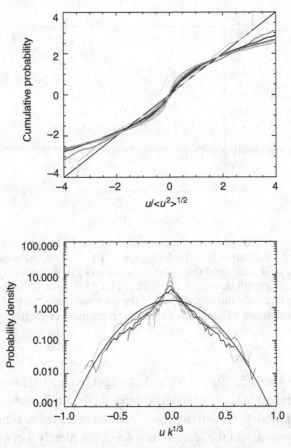

Figure 7.3 Results from a simulation of the standard 3D Sabra model with 30 shells and the parameters $f_n = (1+i)\delta_{n,4}, \nu = 10^{-10}$. The top panel shows the pooled cumulative probabilities of the real and imaginary parts of the shell velocities on a scale in which a Gaussian distribution follows a straight line (probability paper plot). The velocities are normalized such that they have unit variance. The lower panel shows the corresponding probability densities. Here the velocities are scaled with a factor $k^{1/3}$ in accordance with the K41 scaling. The shells 5, 8, 11, 14, 17, and 20 (light to dark) are plotted.

the real and imaginary parts of the velocities for shells 5, 8, 11, 14, 17, and 20 (light to dark) are shown. In the top panel the variables are normalized such that they have unit variance. The scale in the ordinate is such that a Gaussian distribution falls on a straight line (probability paper plot). The bottom panel shows the probability densities of the shell velocities normalized with $k^{1/3}$ according to the K41 scaling. The solid black parabolic curve is the Gaussian with the same variance as the distribution for u_5. It can be seen that for small numbered shells the distributions are Gaussian, while they become more and more stretched with increasing shell number. By comparing (7.20) and (7.21) we obtain

$$\frac{1}{n}\log\left(\int p_n(u)|u|^p du\right) = -\log\lambda\zeta(p), \qquad (7.22)$$

implying that the left side is independent of n. Scaling in the inertial range implies that the relation between $p_n(u)$ and $p_m(u)$ in some sense is the same as the relation between $p_{n+k}(u)$ and $p_{m+k}(u)$. This is the essence also of the multiplicative processes proposed to describe the phenomenology of anomalous scaling in turbulence. To be a little more specific, we can now make the assumption that for $n > m$ there exists a functional relationship so that

$$p_n(u) = \mathcal{F}_{n-m}[p_m(u)]. \qquad (7.23)$$

According to the scaling assumption, the functional $\mathcal{F}_{n-m}$ can only depend on the difference $n - m$. This is the Markov chain assumption. It follows that all the PDFs are now determined by the first PDF, $p_0(u)$:

$$p_n(u) = \mathcal{F}\circ\cdots\circ\mathcal{F}[p_0(u)], \qquad (7.24)$$

where $\mathcal{F} = \mathcal{F}_1$. Equation (7.22) can now be expressed in terms of $\mathcal{F}$ and $p_0(u)$:

$$\frac{1}{n}\log\left(\int \mathcal{F}\circ\cdots\circ\mathcal{F}[p_0(u)]\,|u|^p du\right) = -\zeta(p)\log\lambda, \qquad (7.25)$$

where $\mathcal{F}$ is applied n times inside the integral. From (7.25), $\mathcal{F}$ can be determined from the requirement that the left side is independent of n. In the limit $n \to \infty$ the actual shape of $p_0(u)$ should not matter and can reasonably be taken as Gaussian. However, if the scaling assumption (7.22) is right, the solution to (7.25) cannot be unique since this would imply that $\zeta(p)$ would be some universal function independent of the specific form of the shell model equations (or the Navier–Stokes equation). This is not the case because numerical investigations show that the scaling exponents depend on the free parameters of the shell model (the ϵ and λ in the GOY model). However, the above approach is independent of the governing equation and relies only on the scaling assumption.

7.6 Problems

7.1 **The Cantor set** in the interval [0, 1] is constructed by removing the interval
]1/3, 2/3[leaving the two intervals [0, 1/3] and [2/3, 1]. From these the middle
third is removed and so on. Assume that the Cantor set is the active set for
turbulence.

Determine the probability, P_l, of finding an active region at the level l.

Determine the fractal dimension of the Cantor set.

Using the β-model, calculate the anomalous scaling exponents, ζ_p, in this
case.

8

Equilibrium statistical mechanics

It is often necessary to describe a dynamical system of many degrees of freedom statistically. This can be due to lack of knowledge of the specific state of the system, or because only average quantities are of importance. An example is the temperature of a gas, being the average kinetic energy of the individual molecules. Since we define a dynamical system through a set of deterministic equations,

$$\dot{x}_i(t) = f_i(\mathbf{x}, t), \tag{8.1}$$

with some specified initial conditions $\mathbf{x}(0) = \mathbf{x}_0$, we must specify what we mean by a statistical description. Some of the basic questions, such as the conditions for ergodicity, are still not completely answered in the case of dynamical systems. Here, without too much rigor, we will state a few basic results.

8.1 The statistical ensemble

A way of giving a statistical description of a dynamical system is by specifying that $\mathbf{x}_0$ is a stochastic initial condition drawn from some specified distribution $\psi_0(\mathbf{x})$. The state of the system at a given time t is determined from (8.1) through the operator

$$\tilde{\mathbf{f}}(\mathbf{x}_0, t) = \mathbf{x}(t), \tag{8.2}$$

where

$$\mathbf{x}(t) = \mathbf{x}_0 + \int_0^t \mathbf{f}[\mathbf{x}(\tau)]d\tau. \tag{8.3}$$

Then we can define a statistical quantity as

$$\langle g(\mathbf{x}) \rangle(t) = \int \psi_0(\mathbf{x}) g[\tilde{\mathbf{f}}(\mathbf{x}_0, t)]d\mathbf{x}. \tag{8.4}$$

The probability density for $\mathbf{x}(t)$ is given by the usual rule for transformations of probabilities as

$$\psi_t(\mathbf{x}) = \frac{\psi_0[\tilde{\mathbf{f}}^{-1}(\mathbf{x},t)]}{\left|\partial_j \tilde{f}_i [\tilde{\mathbf{f}}^{-1}(\mathbf{x},t)])\right|}, \tag{8.5}$$

where $|\partial_j \tilde{f}_i|$ is the determinant of the Jacobian matrix of $\tilde{\mathbf{f}}$. From this we can rewrite (8.4) as

$$\langle g(\mathbf{x})\rangle(t) = \int \psi_t(\mathbf{x}) g(\mathbf{x}) d\mathbf{x}, \tag{8.6}$$

which is an ensemble average of $g(\mathbf{x})$.

The ensemble average defined like (8.6) depends on both t and the initial distribution $\psi_0(\mathbf{x})$. The dependence on t can only be eliminated when (8.1) is an autonomous set of equations, as obtained if f_i does not depend explicitly on time in (8.1). This is the same as assuming translational invariance with respect to time: $\mathbf{x}_{t+s} = \tilde{\mathbf{f}}(\mathbf{x}_0, t+s) = \tilde{\mathbf{f}}[\tilde{\mathbf{f}}(\mathbf{x}_0,s),t] = \tilde{\mathbf{f}}(\mathbf{x}_s,t)$. The dependence of the initial distribution vanishes if ψ_t has a limit ψ_∞ as $t \to \infty$, where ψ_∞ is independent of ψ_0. The limiting distribution ψ_∞ is then called the *invariant measure* with respect to the dynamics. If, after some time t and independent of ψ_0, the measure $\psi_\tau \approx \psi_\infty$ for $\tau > t$, then the mean (8.6) can be obtained as the temporal average

$$\langle g(\mathbf{x})\rangle = \lim_{T \to \infty} \frac{1}{T} \int_\tau^{\tau+T} g[x(s)] ds \tag{8.7}$$

over a period $T > t$. If this is the case the system is said to be *ergodic*. The ergodicity hypothesis for a dynamical system can be stated in terms of the phase space flow. For a Hamiltonian system, the phase space flow will be incompressible (Liouville's theorem) on the constant energy surface. The ergodicity hypothesis then states that all infinitesimal areas on the energy surface will be visited equally often and the invariant measure is the Lebesgue measure of the energy surface. A simple example shows that this is, however, not always the case. Consider a stiff pendulum with length l rotating freely around a horizontal axis. The dynamics of the pendulum is governed by Newton's equation: $l\ddot{\theta} = -g\sin\theta$. The evolution $\theta(t)$ cannot easily be found, but the phase space portrait $(\theta, \dot{\theta})$ is readily derived from energy conservation. By direct inspection it is seen that $E = gl(1 - \cos\theta) + l^2\dot{\theta}^2$ (sum of potential and kinetic energy) is conserved. Solving for $\dot{\theta}$: $\dot{\theta} = \pm\sqrt{2(E - gl(1 - \cos\theta))}$, the phase space flow is shown for different energies in Figure 8.1. The heavy upper curve corresponds to counter-clockwise rotation while the heavy lower curve corresponds to clockwise rotation. Which of the two disconnected branches is realized depends on the initial condition. In this simple example, the ensemble average $\langle \dot{\theta} \rangle$ will be zero, while the temporal average will, depending on the initial condition, be either positive or negative. Thus the ergodicity hypothesis does not hold in this case.

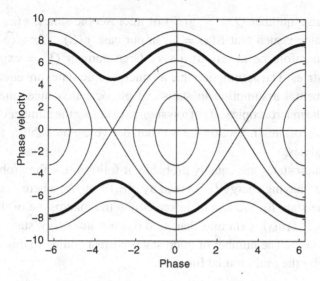

Figure 8.1 Phase space for a stiff pendulum rotating around a horizontal axis. For energies $E < 2gl$, the phase space is a single closed loop. For $E > 2gl$, the phase space divides into two separate curves (heavy lines) corresponding to counterclockwise and clockwise rotations, respectively.

In a dissipative system the phase space flow is convergent and contracts into an attractor with zero Lebesgue measure. Such systems may have disconnected attractors. A given initial point in phase space will then asymptotically end up on one attractor, and the initial point is then said to lie in the basin of attraction for the attractor. Such a dissipative system obviously can only be ergodic if it has exactly one attractor. The Navier–Stokes equation and shell models are believed, but not proven, to be of this kind.

8.2 The partition function

Let us consider a conservative dynamical system with a large number of degrees of freedom (d.o.f.) N governed by (8.1). The total energy of the system is a conserved quantity, $E_{\text{total}} = \sum_n |x_n|^2 = \sum_n E_n$. We are now interested in the probability $p(E)$ for a given d.o.f. x_n to have a given energy $E = |x_n|^2$. For a quantum mechanical system there will in general be a set of discrete energy levels $\mathcal{E}_i$ with $i = 1, \ldots$ for each of the N d.o.f.

In the case of a classical system, like a fluid, it is convenient to separate the energy into a discrete set of energy levels as well. Each level E_i covers the energy range $(i-1)\delta E < E < i\delta E$ with $i = 1, \ldots, M$ and $M\delta E = E_{\text{total}}$.

Assume that the microscopic state of the system of N d.o.f. is described by $(x_1, x_2, \ldots, x_N)$, with energy $E_{\text{total}} = \sum_n E_n = \sum_n x_n^2$. Each of the d.o.f. can be in any of a large number $Q = \sum_i g(E_i)$ of microscopic states; $g(E_i)$ denotes the number of states x such that $E(x) = E_i$, in our case $g(E) \sim 2\pi \sqrt{(E)}$. The entire system can therefore be in any of a very large number $Q^N = \exp(N \log Q)$ of microscopic states. The assumption that all microscopic states are equally probable is the fundamental assumption in statistical physics. It reduces the problem of determining the macroscopic state of a system to counting the number of microstates for a given macrostate. In Section 8.3 we shall derive the probability from a purely geometric analysis.

Since all microstates are equally probable, it follows that the probability $p(E_i)$ of a single d.o.f having energy E_i is given by $p(E_i) = n_i/N$, where n_i is the number of d.o.f. in the state E_i. We can therefore define the macrostate of the system by the vector $(n_1, \ldots, n_M)$, denoting that n_1 d.o.f. are in energy state E_1, etc., and $n_1 + \ldots + n_M = N$. The number of microstates corresponding to this macrostate is simply given by the multinomial function,

$$\frac{N!}{n_1! \ldots n_M!} = \exp(-N \sum_i \frac{n_i}{N} \log \frac{n_i}{N}) = \exp(-N \sum_i p(E_i) \log(p_i)), \qquad (8.8)$$

where we have used Stirlings formula $\log(n!) \approx n \log n - n$, for the first equality. The most probable state of the system is the state which maximizes the exponent (N is constant):

$$S \equiv \langle -\log p \rangle = -\sum_{i=1}^{M} p(E_i) \log p(E_i) = -\int p(E(x)) \log p(E(x)) \mathrm{d}x, \qquad (8.9)$$

where the integral is over the phase space of a single d.o.f. and obtained in the continuum limit $\delta E \to 0$. We can define S as the entropy, and we will denote the state of maximum entropy as the equilibrium state. From (8.8) we define the entropy of the whole system to be $S_{\text{total}} = NS$, making the entropy an extensive (proportional to N) variable, as it should be. Note that for a real flow the entropy of the molecular state of the fluid and the entropy generated in the fluid by viscosity are many orders of magnitude larger than the entropy defined here. The molecular entropy is irrelevant for describing the flow, except perhaps for the case where the temperature field and heat exchange are important.

The maximization of S must be performed with the constraint

$$\int p(E) \, \mathrm{d}E = 1, \qquad (8.10)$$

since p is a probability. This is done using the technique of Lagrange multipliers. The integration (8.9) must be performed over all possible microscopic states, under the further constraint that the total energy of the whole system is constant,

$$\langle E \rangle = \int E \, p(E) dx = \text{constant}. \qquad (8.11)$$

This constraint is introduced through another Lagrange multiplier. With the two Lagrange multipliers for the two constraints and using the variation operator δ, we obtain for the maximum of (8.9) that

$$\delta \int (-p \log p - \lambda p - \beta E p) dx$$

$$= \int (-\log p - 1 - \lambda - \beta E) \delta p \, dx = 0, \qquad (8.12)$$

where the integral is over all possible microscopic states and where the parenthesis in the last integrand must be zero because δp is an arbitrary variation of p. This gives

$$p(E) = \exp[-1 - \lambda - \beta E] = \exp(-\beta E)/Z(\beta), \qquad (8.13)$$

where we have defined $Z(\beta) = \exp(1 + \lambda)$. This probability is the Boltzmann factor. The argument β for $Z(\beta)$ comes about since λ must be determined from the normalization of the probability, giving

$$Z(\beta) = \int \exp(-\beta E) \, dx. \qquad (8.14)$$

The function $Z(\beta)$ is the partition function and is the single most important expression in equilibrium statistical mechanics. This should not be confused with the enstrophy, which is denoted by the same symbol. Most of equilibrium statistical mechanics, including the study of phase transitions, deals with calculating the partition function. The reason for this is that any statistical quantity $\langle f(E) \rangle$, can in principle, be calculated from the partition function. This follows from the calculation

$$\frac{(-1)^n}{Z(\beta)} \frac{\partial^n Z(\beta)}{\partial \beta^n} = \frac{1}{Z(\beta)} \int E^n \exp(-\beta E) \, dx = \langle E^n \rangle. \qquad (8.15)$$

Any function $\langle f(E) \rangle$ is then obtained as a Taylor expansion in powers $\langle E^n \rangle$. As an example we can calculate the mean energy of an ideal gas of N molecules, where each molecule i has energy $E_i = (x_i^2 + y_i^2 + z_i^2)/2$:

$$Z(\beta) = \int \exp\left(-\beta \sum_{i=1}^{N} E_i\right) (\Pi dx)$$

$$= \left(\int \exp(-\beta E) dx dy dz\right)^N = \left(\frac{\pi}{\beta}\right)^{3N/2};$$

$$\frac{\partial Z(\beta)}{\partial \beta} = -\sum_{j=1}^{N} \int E_j \exp\left(-\beta \sum_{i=1}^{N} E_i\right)(\Pi dx)$$

$$= -\sum_{j=1}^{N} \int E_j \exp(-\beta E_j) \int \prod_{i\neq j}^{N} \exp(-\beta E_i)(\Pi dx)$$

$$= -N \int E \exp(-\beta E) dx dy dz \left(\int \exp(-\beta E) dx dy dz\right)^{N-1}$$

$$= -N \frac{3}{2\beta}\left(\sqrt{\frac{\pi}{\beta}}\right)^3 \left(\frac{\pi}{\beta}\right)^{3(N-1)/2}; \tag{8.16}$$

and thus according to (8.15):

$$\langle E \rangle = -\frac{1}{Z(\beta)}\frac{\partial Z(\beta)}{\partial \beta} = \frac{3}{2}N\beta^{-1}. \tag{8.17}$$

This is the well known result that β^{-1} is identified as the temperature kT (where k is the Boltzmann constant), and the mean (kinetic) energy per molecule is $3kT/2$ or $kT/2$ per degree of freedom.

We see that we have arrived at the second law of thermodynamics through a back door: for two subsystems in statistical equilibrium the mean kinetic energy per molecule is the same. The variance of the energy of the subsystem will be inversely proportional to the number of molecules N, which vanish in the thermodynamic limit $N \to \infty$, and the subsystems will have the same temperature β^{-1}. Energy transfer from a low temperature region to a high temperature region is very unlikely and thus prohibited in the thermodynamic limit. Use of the partition function in statistical mechanics is very similar to use of the characteristic function in statistics and the generating function in field theory, which we shall discuss briefly later.

The partition function can be generalized to the situation where not only the energy but other quantities also are given through a conservation law. For each conserved quantity $E_m(x)$ with

$$\langle E_m \rangle = \int E_m(x) p[E_m(x)] dx = \text{constant} \tag{8.18}$$

we have a Lagrange multiplier β_m, and by the procedure (8.12)–(8.14) the partition function becomes

$$Z(\beta_1, \beta_2, \ldots, \beta_n) = \int \exp(-\beta_1 E_1 - \beta_2 E_2 - \cdots - \beta_n E_n) dx. \tag{8.19}$$

In this case there are n generalized temperatures, $\beta_1^{-1}, \ldots, \beta_n^{-1}$, signifying the mean values of $E_1, \ldots, E_n$, respectively. This corresponds to the structure function in the theory of statistics if $E_m = x^m$ is the mth moment of the variable x.

8.3 Phase space geometry

A system in statistical equilibrium has the same temperature everywhere and will thus exhibit an equipartitioning (equal average amount) of the energy between the degrees of freedom of the system. In a field, like the velocity field in a fluid, the concept of degrees of freedom is perhaps not completely straightforward. Obviously a continuous flow field has infinite dimensions, however, since the field is smooth due to viscosity below the Kolmogorov scale η, the field is well described in a grid with lattice constant η so that the number of degrees of freedom is about $n = (L/\eta)^d$, where L is the integral scale and d is the dimension of space. As a corollary from (1.11) it follows that the number of degrees of freedom increases with Reynolds number as $n \sim \mathrm{Re}^{3d/4}$. The same number of degrees of freedom is obtained on a basis of plane waves of wave vector $\mathbf{k}$ with coordinates in the range $2\pi/L < k_i < 2\pi/\eta$. The constant mean energy per degree of freedom does not depend on which basis we count the degrees of freedom. This can be understood by expanding the field in terms of an orthonormal basis:

$$\mathbf{u} = \sum_{i=1}^{n} \langle \mathbf{u} | \mathbf{e}_i \rangle \mathbf{e}_i. \tag{8.20}$$

Here we have used the quantum mechanical "bracket notation" for the inner products. The total energy is conserved, and from the Parseval identity we have

$$2E = \int u(\mathbf{x})^2 d\mathbf{x} = \sum_i |\langle \mathbf{u} | \mathbf{e}_i \rangle|^2. \tag{8.21}$$

Equipartition of the energy implies that the statistical ensemble of states is evenly distributed on the energy sphere defined by (8.21) in the phase space spanned by $\mathbf{e}_i$. The sphere is invariant with respect to any rotation of the basis $\mathbf{e}_i \to \mathbf{e}_i'$ and the mean energy per degree of freedom in the rotated basis $\mathbf{e}_i'$ is $1/\beta = kT$ as well.

From this kind of geometric consideration we can re-derive the fundamental statistical distributions of a conservative system of N degrees of freedom. The probability distribution function $\phi_N(u_i)$ for the individual degrees of freedom u_i is obtained by assuming ergodicity. This implies uniform probability on constant energy surfaces in phase space. These surfaces are the surfaces of the N dimensional spheres defined by

$$2E \equiv NkT = \sum_{i=1}^{N} u_i^2. \tag{8.22}$$

Then $\phi_N(u)$ is given by

$$\phi_N(u) = \frac{\Omega_{N-1}\left(\sqrt{r^2 - u^2}\right)}{\Omega_N(r)} \frac{r}{\sqrt{r^2 - u^2}}, \tag{8.23}$$

where $r = \sqrt{NkT}$, and $\Omega_N(r) = 2\pi^{N/2} r^{N-1}/\Gamma(N/2)$ is the area of the surface of the N-dimensional sphere with radius r. The last factor on the right side comes from projecting the $N - 1$ dimensional sphere onto the u-axis. Neglecting the trivial normalization, we thus have

$$\phi_N(x) \sim (NkT - x^2)^{(N-3)/2} = (NkT)^{(N-3)/2} \left(1 - \frac{x^2}{NkT}\right)^{(N-3)/2}. \tag{8.24}$$

For many degrees of freedom, we get

$$\phi_N(x) \sim \left(1 - \frac{x^2}{NkT}\right)^{(N-3)/2} = \exp\left[\frac{N-3}{2} \log\left(1 - \frac{x^2}{NkT}\right)\right]$$

$$\rightarrow_{N \to \infty} \exp\left(-\frac{x^2}{2kT}\right) = \exp\left(-\frac{E_x}{kT}\right). \tag{8.25}$$

This is the Boltzmann distribution (8.13) derived purely from ergodicity and the phase space geometry; the distribution is the Gaussian distribution for the velocities.

8.4 Statistical equilibrium and turbulence

In equilibrium a turbulent fluid should have the energy evenly distributed, or equipartitioned, between the degrees of freedom of the flow. The number of degrees of freedom is homogeneous in **k**-space and thus proportional to the volume. The number of degrees of freedom with modulus of the wave vector between k and $k + dk$ is then proportional to dk times the area of the sphere with radius k. Thus the energy spectrum scales as $E(k) \sim k^{d-1}$ where d is the dimension of space. For 3D turbulence this is $E(k) \sim k^2$, in marked contrast to the observed $K41$ spectrum $E(k) \sim k^{-5/3}$. In the 2D case there is a forward cascade of enstrophy corresponding to the enstrophy spectrum $Z(k) = k^2 E(k) \sim k^{-1}$. The equipartition of enstrophy implies $Z(k) \sim k$, again in conflict with observations. In the range of wave numbers smaller than the scale of the force, the energy spectrum is $E(k) \sim k^{-5/3}$ for an inverse energy cascade (cascade towards small k), while the spectrum for equipartition is $E(k) \sim k$.

To gain insight into the conditions under which the turbulent cascade can be viewed as a statistical equilibrium, we return to the case of shell models. In a statistical equilibrium the shell velocities have a probability distribution governed by an equipartitioning of the inviscid invariants between the shells. For the 2D shell model defined according to (3.30), case $\epsilon > 1$, there are two positive conserved quantities, energy $E^{(1)}$ and enstrophy $E^{(2)}$. Using (8.19), the probability density for the nth shell is

$$P_n(u_n) \sim \exp(-BE_n^1 - AE_n^2) = \exp(-Bk_n^{\alpha_1}|u_n|^2 - Ak_n^{\alpha_2}|u_n|^2), \qquad (8.26)$$

in accordance with the notation introduced in Section 3.6. Thus the temporal mean of any function g of the shell velocities is given as

$$\langle g \rangle = \frac{\int g(u_1, \ldots, u_N) \prod_n \exp(-BE_n^1 - AE_n^2) \, du_n}{\int \prod_n \exp(-BE_n^1 - AE_n^2) \, du_n}, \qquad (8.27)$$

where A and B are Lagrange multipliers that reflect the conservation of energy and enstrophy when maximizing the entropy of the system. These multipliers correspond to inverse temperatures and are denoted as inverse *energy* and *enstrophy temperatures* respectively (Kraichnan & Montgomery, 1980).

The shell velocities themselves will be independent, and Gaussian random variables with variance $\sigma(u_n)^2 = (Bk_n^{\alpha_1} + Ak_n^{\alpha_2})^{-1}$. The average values of the energy and enstrophy become

$$\langle E_n^1 \rangle = \frac{1}{2}k_n^{\alpha_1} \langle |u_n|^2 \rangle = \frac{1}{2}(B + Ak_n^{\alpha_2 - \alpha_1})^{-1},$$

$$\langle E_n^2 \rangle = \frac{1}{2}k_n^{\alpha_2} \langle |u_n|^2 \rangle = \frac{1}{2}(Bk_n^{\alpha_1 - \alpha_2} + A)^{-1}. \qquad (8.28)$$

For $1 < \epsilon < 2$ we have from (3.34) that $\alpha_1 = 0 < \alpha_2 = -\log_\lambda(\epsilon - 1)$, with $k_n = \lambda^n$. Then for $k_n \to 0$ ($n \to -\infty$) we have equipartitioning of energy, $k_n^{\alpha_1} \langle |u_n|^2 \rangle = B^{-1}$ and the scaling $|u_n| \sim k_n^{-\alpha_1/2}$. For the other branch, $k \to \infty$, we have equipartitioning of enstrophy, $k_n^{\alpha_2} \langle |u_n|^2 \rangle = A^{-1}$ and the scaling $|u_n| \sim k_n^{-\alpha_2/2}$.

In the case of no force and no viscosity the equilibrium will depend on the ratio A/B between the initial temperatures A^{-1}, B^{-1}. To illustrate this, Figure 8.2 shows the spectra from a simulation of the GOY model without force and viscosity, but with two different initial spectral slopes of the velocity field. The larger the slope the higher the ratio of the energy temperature to the enstrophy temperature. The figure shows the equilibrium spectra for $\epsilon = 5/4$ and $\nu = f = 0$ in the cases of initial slopes -1 and -0.8. The full lines are the equilibrium distribution given by (8.28) for $A/B = 10^2$ and $A/B = 10^{-2}$, respectively.

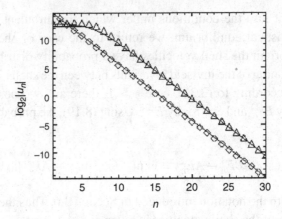

Figure 8.2 The mean value of the shell velocities as a function of shell number on a logarithmic scale (base λ), for the 2D case: $\epsilon = 5/4$, $k_0 = \lambda^{-4}$, $\lambda = 2$, $n = 30$, $v = f = 0$. Diamonds correspond to an initial spectral slope of -1.0, which is a high value of A/B. The corresponding curve is the calculated statistical equilibrium distribution for $A/B = 10^2$. Triangles correspond to an initial spectral slope of -0.8, which is a lower value of A/B. The curve is the calculated statistical equilibrium distribution for $A/B = 10^{-2}$.

8.5 Cascade or equilibrium

The spectral slope γ is defined from the scaling $|u| \sim k^\gamma$. For the forward (towards large k) enstrophy cascade the slope is given by (4.2) as $\gamma_{\text{cascade}} = -(\alpha + 1)/3$. Assuming statistical equilibrium, the enstrophy equipartitioning branch has the spectral slope $\gamma_{\text{equilibrium}} = -\alpha/2$. Thus for the 2D case where $\alpha = 2$ we have $\gamma_{\text{cascade}} = \gamma_{\text{equilibrium}}$ and we cannot distinguish between statistical quasi-equilibrium and a cascade. This was pointed out by Aurell *et al.* (1994) and it was argued that the model can be described as being in statistical quasi-equilibrium with the enstrophy transfer described as an effective diffusion and not an enstrophy cascade. This coinciding scaling is a feature of the shell model not present in the real 2D flow, where the statistical equilibrium energy spectrum scales as k^{-1} and the cascade energy spectrum scales as k^{-3}.

For other values of α the scaling of the two cases is different. The linear relations between α and γ in the case of equilibrium or in the case of cascade are shown in Figure 8.3. The first axis is the parameter α where the enstrophy is defined as $E_n^2 = k_n^\alpha |u_n|^2$. The second axis is the scaling exponent γ. The two curves are the scaling exponent for the enstrophy cascade and enstrophy equilibrium as indicated. The two lines cross at $\alpha = 2$. In the limit $\alpha \to 0$, $\epsilon \to 2$, the energy and the enstrophy become identical and there is only one conserved quantity. In the limit $\alpha \to \infty$ the case $\epsilon = 1$ is approached from above. Which spectral slope will be realized in the shell model is determined from simulations for different values of α.

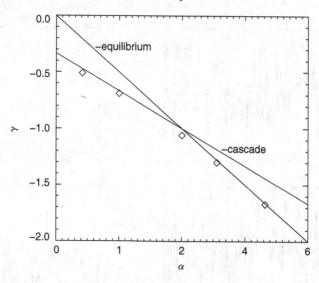

Figure 8.3 The spectral slopes γ calculated from a set of simulations of the 2D GOY model for varying values of ϵ. The parameter α is related to ϵ as $\epsilon - 1 = \lambda^{-\alpha}$, $\lambda = 2$.

For 2D turbulence-like models the scaling slope is everywhere on or slightly below both the cascade and the equilibrium slopes (diamonds in the figure). The classical argument for a cascade is that, given an initial state with enstrophy concentrated at the low wavenumber end of the spectrum, the enstrophy will flow into the high wave numbers in order to establish statistical equilibrium. The ultraviolet catastrophe is then prevented by dissipation in the viscous subrange. Therefore we cannot have a nonequilibrium distribution with more enstrophy in the high wave number part of the spectrum than prescribed by statistical equilibrium. If this was the case, the enstrophy would flow backward from high wave numbers to low wavenumbers and thereby establish equilibrium. This means that the spectral slope in the inertial subrange is always below the slope corresponding to equilibrium. Consequently, the 2D model with $\epsilon = 5/4$ separates two regimes: $1 < \epsilon < 5/4$ where enstrophy equilibrium is achieved, and $5/4 < \epsilon < 2$ where there is an enstrophy cascade through the inertial range.

The two regimes corresponding to equipartitioning and cascade can be understood in terms of timescales for the dynamics of the shell velocities. A rough estimate of the timescales for a given shell n is, from (3.22), given as $T_n \sim (k_n u_n)^{-1} \sim k_n^{-\gamma-1}$. Again $\alpha = 2$ or $\epsilon = 5/4$, corresponding to $\gamma = -1$, separates two regimes. For $\epsilon = 5/4$ the timescale is independent of shell number. For $\epsilon < 5/4$ the timescale increases with n and the fast timescales for small n can equilibrate the enstrophy among the degrees of freedom of the system before the dissipation at the "slow"

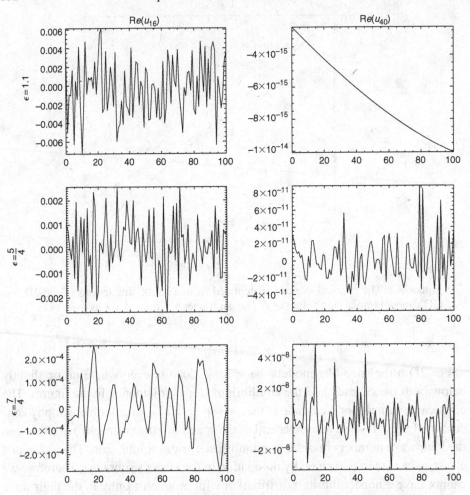

Figure 8.4 Time evolution of shell velocities in the beginning left and end right of the inertial subrange. The typical timescale of shell n scales as $T_n \sim (k_n|u_n|)^{-1} \sim k_n^{-\gamma-1}$. For $\epsilon = 5/4$ the timescale is the same for all shells.

shells has time to be active. Therefore these models exhibit statistical equilibrium. For $\epsilon > 5/4$ the situation is reversed and the models exhibit enstrophy cascades. Time evolutions of the shell velocities from numerical realizations of (3.22) are shown in Figure 8.4. The left hand side shows the evolution of a shell in the beginning of the inertial subrange and the right hand side shows the evolution of a shell at the end of the inertial subrange. The timescale corresponding to the eddy turnover time is the typical time between consecutive maxima or minima of the curves. For $\epsilon = 1.1$ the timescale for shell number 16 is of the order 100/30 (arbitrary units), while the timescale for shell number 40 is much longer than 100. For $\epsilon = 5/4$ the timescales for shell number 16 and shell number 40 are the same, around 30/100, while for $\epsilon = 7/4$ the timescale for shell number 16 is about 100/10 and for shell number 40 it is about 100/30. It has not been possible to make simulations that show an inverse cascade of energy similar to the inverse cascade of energy in 2D NSE turbulence. The dependence of the timescale on the shell number could explain why no inverse cascade branch has been seen in the shell model. At the small wave number end of the spectrum the timescales are long in comparison to the timescales of the inertial range of inverse cascades. Thus there may be sufficient time for dissipation or drag to remove the energy and give time to reach statistical equilibrium.

8.6 Problems

8.1 **Maximum entropy distribution.** Consider probability distributions $p(x)$ on the unit interval $[0, 1]$. Show, using the method of Lagrange multipliers, that the uniform distribution maximizes the entropy $S = \int p \log p \, dx$.

8.2 **Information and entropy.** Observing that a system in statistical equilibrium is in some macroscopic state S (specified by, say, the total energy), we might pose the question of how much information is needed to identify the microscopic state of the system. Following the definition by Shannon (1948), the information I is defined as the number of bits needed to specify the microscopic state, knowing only the macroscopic state.

Assume that the phase space has volume V, corresponding to macroscopic states defined by the total energy in the interval $[E, E + \delta E]$. Assume further that we split the volume into $N = 2^m$ numbered sub-volumes, $N\delta V = V$. Determine how many "yes" or "no" questions are needed in order to specify which one of the sub-volumes contains the microstate of the system.

If the entropy of the system is defined as the logarithm of the number of microstates corresponding to the macrostate, what is the relation between the entropy and the information measure?

Assuming that the energy is fixed (micro-canonical ensemble), and each sub-volume i is equally probable, $p_i = 1/N$, what is the relationship between entropy and probability?

For a system exchanging energy with the surroundings (canonical ensemble), each macrostate E_i has some probability $p(E_i)$. What is the average information needed to completely specify the microstate of the system? Compare the result with equation (8.9).

8.3 **The characteristic function** $K(z)$ for a stochastic (random) variable X is the mean of the harmonics, which is the same as the Fourier transform of the probability density $p(x)$:

$$K(z) = \langle e^{izx} \rangle = \int e^{izx} p(x) \mathrm{d}x.$$

The characteristic function for the Gaussian $p(x) = N \exp[-x^2/(2\sigma^2)]$ is $K(z) = \exp(-\sigma^2 x^2/2)$; $N = 1/\sqrt{2\pi\sigma^2}$ is the normalization.

Consider the sum of two independent centered Gaussian variables $X = X_1 + X_2$, where X_1 has variance σ_1^2 and X_2 has variance σ_2^2. Determine the probability density for X using the fact that the characteristic function for the sum is given by

$$K(z) = K_1(z)K_2(z). \qquad (8.29)$$

Show that the same result is obtained by the usual convolution:

$$p(x) = p_1 * p_2(x) = \int p_1(x_1)p_2(x - x_1)\mathrm{d}x_1,$$

where $p_i(x) = N_i \exp(-x^2/(2\sigma_i^2))$, $i = 1, 2$.

By iteration, show that the sum of n i.i.d. (independent identically distributed) Gaussian variables with variance σ^2 has variance $(\sqrt{n}\sigma)^2$.

Prove Equation (8.29).

Appendix

A.1 Velocity Fourier transforms in 3D

All integral operations are based on results justified in mathematical distribution theory, in particular using the Dirac delta function $\delta(\mathbf{x})$ with the formal Fourier transforms:

$$\mathcal{F}_-[\delta(\mathbf{x})] = 1: \qquad 1 = \int e^{-\iota \mathbf{k}\mathbf{x}}\delta(\mathbf{x})d\mathbf{x}, \qquad (A.1)$$

$$\mathcal{F}_+[1] = \delta(\mathbf{x}): \qquad \delta(\mathbf{x}) = \frac{1}{(2\pi)^3}\int e^{\iota \mathbf{k}\mathbf{x}}d\mathbf{k}, \qquad (A.2)$$

where ι is the imaginary unit. The integral sign $\int$ here and in the following is short for integration over the entire real space $\mathbb{R}^3$ using Cauchy's definition of primary integral value whenever needed. Problems of convergence of the integrals are not treated in this book. Reference is made to the numerous books on mathematical analysis and Fourier theory. In fact, besides using primary integral values, one can simply adopt a formalism by which divergent integrals are limits as $c \to 0$ of convergent integrals defined with the factor $\psi(\mathbf{x};c^2) = \exp(-c^2|\mathbf{x}|^2)$ applied to the integrand. It is also assumed that integral order in multiple integrals can be interchanged without problems (Fubini's theorem) and that integration and differentiation can be interchanged.

A velocity field is written as $\mathbf{u}(\mathbf{x},t)$ where $\mathbf{x}$ (or x_i) is the position vector and t is the time. Most often the time is suppressed in the notation writing $\mathbf{u}(\mathbf{x})$, or, using Cartesian tensor notation, $u_i(\mathbf{x})$. The summation convention applies in the sense that two repeated indices signifies summation over all three dimensions. The Fourier transforms for the velocity field are:

$$\mathcal{F}_-[u_i(\mathbf{x})] = u_i(\mathbf{k}): \qquad u_i(\mathbf{k}) = \frac{1}{(2\pi)^3}\int e^{-\iota \mathbf{k}\mathbf{x}}u_i(\mathbf{x})d\mathbf{x}, \qquad (A.3)$$

$$\mathcal{F}_+[u_i(\mathbf{k})] = u_i(\mathbf{x}): \qquad u_i(\mathbf{x}) = \int e^{\iota \mathbf{k}\mathbf{x}}u_i(\mathbf{k})d\mathbf{k}, \qquad (A.4)$$

where the last transform follows from the first by use of (A.1) and (A.2):

$$\frac{1}{(2\pi)^3} \int u_i(\mathbf{x}) e^{-\iota \mathbf{k}'\mathbf{x}} d\mathbf{x} = \frac{1}{(2\pi)^3} \int \int e^{\iota \mathbf{k}\mathbf{x}} u_i(\mathbf{k}) e^{-\iota \mathbf{k}'\mathbf{x}} d\mathbf{k} d\mathbf{x}$$

$$= \int u_i(\mathbf{k}) d\mathbf{k} \frac{1}{(2\pi)^3} \int e^{\tau(\mathbf{k}-\mathbf{k}')\mathbf{x}} d\mathbf{x}$$

$$= \int u_i(\mathbf{k}) \delta(\mathbf{k} - \mathbf{k}') d\mathbf{k} = u_i(\mathbf{k}'). \qquad (A.5)$$

The property of the delta function as a substitution operator is used in the last step.

The notation for the Fourier transforms is kept minimal in the sense that the distinction is read from the argument symbols. This convention requires that $\mathbf{x}, \mathbf{x}'$, etc., are used as standard for the position vectors and $\mathbf{k}, \mathbf{k}'$, etc., are used as standard for the *wave number vectors* (short: *wave vectors*). Thus $u_i(\mathbf{x})$ is the velocity field in the geometric space while $u_i(\mathbf{k})$ is the corresponding wave amplitude in the spectral space. The spectral amplitudes are also called the *spectral velocities*. When needed, the spectral velocities are written as $\hat{u}_i$.

The Fourier transform of $u_i u_j$ is

$$\mathcal{F}_-[u_i u_j] = \frac{1}{(2\pi)^3} \int e^{-\iota \mathbf{k}\mathbf{x}} u_i u_j d\mathbf{x}$$

$$= \frac{1}{(2\pi)^3} \int e^{-\iota \mathbf{k}\mathbf{x}} \int e^{\iota \mathbf{k}''\mathbf{x}} u_i(\mathbf{k}'') d\mathbf{k}'' \int e^{\iota \mathbf{k}'\mathbf{x}} u_j(\mathbf{k}') d\mathbf{k}' d\mathbf{x}$$

$$= \int \int \delta(\mathbf{k}' + \mathbf{k}'' - \mathbf{k}) u_i(\mathbf{k}'') u_j(\mathbf{k}') d\mathbf{k}' d\mathbf{k}''$$

$$= \int u_i(\mathbf{k} - \mathbf{k}') u_j(\mathbf{k}') d\mathbf{k}', \qquad (A.6)$$

which for $\mathbf{k} = (0,0,0)$ and contraction gives

$$\frac{1}{(2\pi)^3} \int u_i(\mathbf{x}) u_i(\mathbf{x}) d\mathbf{x} = \int u_i(-\mathbf{k}') u_i(\mathbf{k}') d\mathbf{k}'$$

$$= \int u_i(\mathbf{k}') u_i^*(\mathbf{k}') d\mathbf{k}', \qquad (A.7)$$

noting that since $u_i(\mathbf{x})$ is real, then $u_i(-\mathbf{k}') = u_i^*(\mathbf{k}')$ where * indicates complex conjugation. This is known as *Parseval's identity*.

Replacing $u_i(\mathbf{x})$ and $u_j(\mathbf{x})$ in (A.6) by arbitrary functions $F(\mathbf{x})$ and $G(\mathbf{x})$, respectively, we get the general transformation formula

$$\mathcal{F}_-[F\,G] = \frac{1}{(2\pi)^3} \int e^{-\iota \mathbf{k}\mathbf{x}} F\,G \mathrm{dx} = \int F(\mathbf{k} - \mathbf{k}')\,G(\mathbf{k}')\mathrm{dk}', \qquad (A.8)$$

denoted as the *convolution rule*.

Partial derivatives with respect to time are written as $\partial_t u_i(\mathbf{x})$ or $\partial_t u_i(\mathbf{k})$ for $\partial u_i(\mathbf{x})/\partial t$, or $\partial u_i(\mathbf{k})/\partial t$, respectively. Only if doubt is excluded, the notation $\partial_t u_i$ is used. Differentiation with respect to the position coordinate x_i does not appear for functions of $\mathbf{k}$, of course, and it is therefore frequently sufficient to write $\partial_j u_i$ for $\partial u_i/\partial x_j$.

The Fourier transform of $\partial_j F(\mathbf{x})$ is directly obtained as

$$\mathcal{F}_-[\partial_j F(\mathbf{x})] = \iota k_j F(\mathbf{k}), \qquad (A.9)$$

which is generalized directly to higher order partial derivatives, that is, for each differentiation ∂_i the Fourier transform is multiplied by ιk_i.

The Fourier transforms of products like $F\partial_j G$ and $\partial_i F \partial_j G$ are obtained from the convolution rule as

$$\mathcal{F}_-[F\partial_j G] = \begin{cases} \iota \int F(\mathbf{k} - \mathbf{k}')k_j' G(\mathbf{k}')\mathrm{dk}' \\ \iota \int F(\mathbf{k}')\,(k_j - k_j')G(\mathbf{k} - \mathbf{k}')\mathrm{dk}' \end{cases}, \qquad (A.10)$$

and

$$\mathcal{F}_-[\partial_- F\partial_j G] = \begin{cases} -\int (k_i - k_i')F(\mathbf{k} - \mathbf{k}')k_j' G(\mathbf{k}')\mathrm{dk}' \\ -\int k_i' F(\mathbf{k}')\,(k_j - k_j')G(\mathbf{k} - \mathbf{k}')\mathrm{dk}' \end{cases}. \qquad (A.11)$$

A.2 Rotation of the velocity field

The curl vector $\operatorname{curl} \mathbf{u}$ (or rotation vector $\operatorname{rot} \mathbf{u}$) of the velocity field $\mathbf{u}$ is defined as the vector product

$$\operatorname{curl}\mathbf{u} = \nabla \times \mathbf{u} = \begin{bmatrix} 0 & -\partial_3 & \partial_2 \\ \partial_3 & 0 & -\partial_1 \\ -\partial_2 & \partial_1 & 0 \end{bmatrix}\begin{bmatrix} u_1 \\ u_2 \\ u_3 \end{bmatrix} = \begin{bmatrix} \partial_2 u_3 - \partial_3 u_2 \\ \partial_3 u_1 - \partial_1 u_3 \\ \partial_1 u_2 - \partial_2 u_1 \end{bmatrix}, \qquad (A.12)$$

or written as a Cartesian tensor

$$\omega_i = \epsilon_{ijl}\partial_j u_l, \qquad (A.13)$$

where

$$\epsilon_{ijl} = \begin{cases} 1 & \text{for } i,j,l = 1,2,3 \text{ or } 2,3,1 \text{ or } 3,1,2 \\ -1 & \text{for } i,j,l = 2,1,3 \text{ or } 1,3,2 \text{ or } 3,2,1, \\ 0 & \text{otherwise} \end{cases} \tag{A.14}$$

is the *permutation symbol*, also called the *Levi–Civita symbol*. It is directly seen that the *curl vector field has zero divergence*

$$\text{div curl } \mathbf{u} = \partial_i \omega_i = 0. \tag{A.15}$$

It follows directly from (A.12) and (A.9) that the Fourier transform of the curl vector is

$$\mathcal{F}_-[\text{curl}\,\mathbf{u}] = \iota\,\mathbf{k} \times \mathbf{u}(\mathbf{k}) = \iota \begin{bmatrix} 0 & -k_3 & k_2 \\ k_3 & 0 & -k_1 \\ -k_2 & k_1 & 0 \end{bmatrix} \begin{bmatrix} u_1(\mathbf{k}) \\ u_2(\mathbf{k}) \\ u_3(\mathbf{k}) \end{bmatrix}, \tag{A.16}$$

or in tensor form

$$\mathcal{F}_-[\omega_i] = \iota\,\epsilon_{ijl}\,k_j u_l(\mathbf{k}). \tag{A.17}$$

A.3 Product rules for Levi–Civita symbols

Occasionally products of Levi–Civita symbols appear in the applications. The product $\epsilon_{ijk}\epsilon_{lmn}$ has 3^6 elements of value -1, 0, or 1, and it is expressible in terms of Kronecker's delta through the determinant formula

$$\epsilon_{ijk}\epsilon_{lmn} = \begin{vmatrix} \delta_{il} & \delta_{im} & \delta_{in} \\ \delta_{jl} & \delta_{jm} & \delta_{jn} \\ \delta_{kl} & \delta_{km} & \delta_{kn} \end{vmatrix}. \tag{A.18}$$

Contraction over the the indices i and l in (A.18) gives

$$\begin{aligned}
\epsilon_{ijk}\epsilon_{imn} &= \begin{vmatrix} \delta_{ii} & \delta_{im} & \delta_{in} \\ \delta_{ji} & \delta_{jm} & \delta_{jn} \\ \delta_{ki} & \delta_{km} & \delta_{kn} \end{vmatrix} \\
&= 3\begin{vmatrix} \delta_{jm} & \delta_{jn} \\ \delta_{km} & \delta_{kn} \end{vmatrix} - \delta_{ji}\begin{vmatrix} \delta_{im} & \delta_{in} \\ \delta_{km} & \delta_{kn} \end{vmatrix} + \delta_{ki}\begin{vmatrix} \delta_{im} & \delta_{in} \\ \delta_{jm} & \delta_{jn} \end{vmatrix} \\
&= 3\begin{vmatrix} \delta_{jm} & \delta_{jn} \\ \delta_{km} & \delta_{kn} \end{vmatrix} - \begin{vmatrix} \delta_{jm} & \delta_{jn} \\ \delta_{km} & \delta_{kn} \end{vmatrix} + \begin{vmatrix} \delta_{km} & \delta_{kn} \\ \delta_{jm} & \delta_{jn} \end{vmatrix} \\
&= \begin{vmatrix} \delta_{jm} & \delta_{jn} \\ \delta_{km} & \delta_{kn} \end{vmatrix} = \delta_{jm}\delta_{kn} - \delta_{jn}\delta_{km}. \tag{A.19}
\end{aligned}$$

Any contraction over two indices like $\epsilon_{ijk}\epsilon_{lmj}$, one from each of the two factors, can be transformed to a contraction over the two first indices by cyclic rotation of the indices to obtain $\epsilon_{jki}\epsilon_{jlm} = \delta_{kl}\delta_{im} - \delta_{km}\delta_{il}$. Contraction over two indices in the same factor is zero, of course.

A.4 Divergence theorem (integration by parts)

After multiplication by u_i and integration over the volume of the flow, the Navier–Stokes equation gives

$$D_t \int \frac{1}{2} u_i u_i \mathrm{dx} = - \int u_i \partial_i p \mathrm{dx} + \nu \int u_i \partial_{jj} u_i \mathrm{dx} + \int u_i f_i \mathrm{dx}. \qquad (A.20)$$

Only the case of an incompressible fluid is considered, implying that the divergence $\mathrm{div}\,\mathbf{u} = \partial_i u_i$ is zero everywhere, i.e.,

$$\partial_i u_i = 0. \qquad (A.21)$$

In spectral space, incompressibility has, according to (A.9), the consequence that $k_i u_i(\mathbf{k}) = 0$, that is, *the spectral velocity* $\mathbf{u}(\mathbf{k})$ *is everywhere orthogonal to the wave vector* $\mathbf{k}$.

Moreover, it is assumed that the boundary conditions are periodic in each of the position coordinates of $\mathbf{x}$. Using (A.21) it is seen that

$$u_i \partial_i p = \partial_i(u_i p) - (\partial_i u_i)p = \partial_i(u_i p) \qquad (A.22)$$

is the divergence of the vector field $u_i p$. Thus Gauss's integration theorem (the divergence theorem of "integration by parts") together with the boundary conditions show that the first term on the right side of (A.20) vanishes. Also observing that

$$\partial_j(u_i \partial_j u_i) = \partial_j u_i \partial_j u_i + u_i \partial_{jj} u_i \qquad (A.23)$$

is the divergence of the vector field $u_i \partial_j u_i$, it follows that

$$0 = \int \partial_j u_i \partial_j u_i \mathrm{dx} + \int u_i \partial_{jj} u_i \mathrm{dx}. \qquad (A.24)$$

As a result the right side of (A.20) reduces to

$$-\nu \int \partial_j u_i \partial_j u_i \mathrm{dx} + F, \qquad (A.25)$$

where $F = \int u_i f_i \mathrm{dx}$.

By comparison with (A.12), it is seen that

$$\omega_i \omega_i = \partial_j u_i \partial_j u_i - \partial_j u_i \partial_i u_j, \qquad (A.26)$$

where the last term can be written as

$$\partial_j u_i \partial_i u_j = \partial_j \partial_i (u_i u_j) - \partial_j u_j \partial_i u_i, \tag{A.27}$$

in which after integration the last term vanishes due to (A.21) and the first term vanishes because it is the divergence of the vector field $\partial_i(u_i u_j)$. Therefore the right side of (A.20) reduces further to give the *energy equation*

$$D_t \int \frac{1}{2} u_i u_i \mathrm{d}\mathbf{x} = -\nu \int \omega^2 \mathrm{d}\mathbf{x} + F, \tag{A.28}$$

where $\omega^2 = \omega_i \omega_i$. The rotation vector $\omega_i(\mathbf{x})$ is called the *vorticity* of the velocity field $\mathbf{u}(\mathbf{x})$.

A.5 The vorticity equation

Application of the operator $\epsilon_{ijl}\partial_j$ to the NSE gives the *vorticity equation*:

$$\begin{aligned}
0 &= \epsilon_{ijl}\partial_j[\partial_t u_l + u_m \partial_m u_l + \partial_l p - \nu \partial_{mm} u_l - f_l] \\
&= \partial_t \omega_i + u_m \partial_m \omega_i + \epsilon_{ijl}\partial_j u_m \partial_m u_l + \epsilon_{ijl}\partial_l \partial_j p - \nu \partial_{mm}\omega_i - \epsilon_{ijl}\partial_j f_l \\
&= \partial_t \omega_i + u_j \partial_j \omega_i - \omega_j \partial_j u_i - \nu \partial_{jj}\omega_i - \epsilon_{ijl}\partial_j f_l, \tag{A.29}
\end{aligned}$$

because $\epsilon_{ijl}\partial_l \partial_j p = 0$ follows from $\epsilon_{ijl} = -\epsilon_{ilj}$, and

$$\epsilon_{ijl}\partial_j u_m \partial_m u_l = -\omega_j \partial_j u_i + \omega_i \partial_j u_j = -\omega_j \partial_j u_i, \tag{A.30}$$

where the incompressibility is used in the last equality. It is sufficient to make a control of (A.30) for $i = 1$:

$$\begin{aligned}
\epsilon_{1jl}\partial_j u_m \partial_m u_l &= \partial_2 u_m \partial_m u_3 - \partial_3 u_m \partial_m u_2 \\
&= \partial_2 u_1 \partial_1 u_3 - \partial_3 u_1 \partial_1 u_2 + (\partial_2 u_3 - \partial_3 u_2)\partial_2 u_2 + (\partial_2 u_3 - \partial_3 u_2)\partial_3 u_3 \\
&= \partial_2 u_1 \partial_1 u_3 - \partial_3 u_1 \partial_1 u_2 - \omega_1 \partial_1 u_1 + \omega_1 \partial_j u_j \\
&= -\omega_1 \partial_1 u_1 - \partial_2 u_1(-\partial_1 u_3 + \partial_3 u_1) - \partial_3 u_1(\partial_1 u_2 - \partial_2 u_1) + \omega_1 \partial_j u_j \\
&= -\omega_j \partial_j u_1 + \omega_1 \partial_j u_j. \tag{A.31}
\end{aligned}$$

The product $\omega_j \partial_j u_i$ is called the *stretching and bending vector* because

$$\omega_j \partial_j u_i = \begin{bmatrix} \omega_1 \partial_1 u_1 \\ \omega_2 \partial_2 u_2 \\ \omega_3 \partial_3 u_3 \end{bmatrix} + \begin{bmatrix} \omega_2 \partial_2 u_1 + \omega_3 \partial_3 u_1 \\ \omega_1 \partial_1 u_2 + \omega_3 \partial_3 u_2 \\ \omega_j \partial_1 u_3 + \omega_2 \partial_2 u_3 \end{bmatrix}, \tag{A.32}$$

where the first column vector is the stretching of ω_i and the second column is the bending of ω_i.

A.6 Helicity

The integral $\int u_i \omega_i \mathrm{d}\mathbf{x}$ of the projection of the velocity vector u_i on the curl vector ω_i (or vice versa) is called the *helicity*. The time derivative of the helicity is

$$\frac{\mathrm{d}H}{\mathrm{d}t} = \int \left(\frac{\mathrm{d}u_i}{\mathrm{d}t} \omega_i + u_i \frac{\mathrm{d}\omega_i}{\mathrm{d}t} \right) \mathrm{d}\mathbf{x}$$

$$= \int (-\omega_i \partial_i p + v \omega_i \partial_{jj} u_i + u_i \omega_j \partial_j u_i + v u_i \partial_{jj} \omega_i) \mathrm{d}\mathbf{x}, \qquad (A.33)$$

obtained by use of the force free NSE and the force free vorticity Equation (A.29). Term by term of the integrand we have

$$\omega_i \partial_i p = \partial_i(\omega_i p) + (\partial_i \omega_i) p = \partial_i(\omega_i p),$$

$$\omega_j u_i \partial_j u_i = \frac{1}{2}[\partial_j(u_i u_i \omega_j) - u_i u_i \partial_j \omega_j] = \frac{1}{2}\partial_j(u_i u_i \omega_j),$$

$$\omega_i \partial_{jj} u_i + u_i \partial_{jj} \omega_i = \partial_j(u_i \partial_j \omega_i + \omega_i \partial_j u_i) - 2\partial_j \omega_i \partial_j u_i$$

$$= \partial_{jj}(u_i \omega_i) - 2\partial_j \omega_i \partial_j u_i, \qquad (A.34)$$

knowing that $\partial_j \omega_j = 0$. Since $\partial_i(\omega_i p)$, $\partial_j(u_i u_i \omega_j)$, and $\partial_{jj}(u_i \omega_i)$ are divergences of the vector fields $\omega_i p$, $u_i u_i \omega_j$, and $\partial_j(u_i \omega_i)$, respectively, the integrals of these terms vanish by Gauss's integration theorem. Thus (A.33) reduces to

$$\frac{\mathrm{d}H}{\mathrm{d}t} = -2v \int \partial_j \omega_i \partial_j u_i \mathrm{d}\mathbf{x}. \qquad (A.35)$$

According to the convolution rule (A.8), the Fourier transform of the product $\partial_j \omega_i \partial_j u_i$ is

$$\mathcal{F}_-[\partial_j \omega_i \partial_j u_i](\mathbf{k}) = -\int k'_j \omega_i(\mathbf{k}')(k_j - k'_j) u_i(\mathbf{k} - \mathbf{k}')\mathrm{d}\mathbf{k}', \qquad (A.36)$$

which for $\mathbf{k} = 0$ gives, see (A.3) and (A.17),

$$\frac{1}{(2\pi)^3} \int \partial_j \omega_i \partial_j u_i \mathrm{d}\mathbf{x} = \int k^2 \omega_i(\mathbf{k}) u_i^*(\mathbf{k})\mathrm{d}\mathbf{k}$$

$$= \iota \epsilon_{ijl} \int k^2 k_j u_l(\mathbf{k}) u_i^*(\mathbf{k})\mathrm{d}\mathbf{k}$$

$$= \iota \int k^2 \mathbf{k} \cdot [\mathbf{u}(\mathbf{k}) \times \mathbf{u}^*(\mathbf{k})]\mathrm{d}\mathbf{k}. \qquad (A.37)$$

A.7 Energy balance between triads

From the inviscid spectral NSE with no force

$$\partial_t u_i(\mathbf{k}) = -\iota k_j \int \left(\delta_{il} - \frac{k_i k_l'}{k^2} \right) u_j(\mathbf{k}') u_l(\mathbf{k} - \mathbf{k}') d\mathbf{k}', \qquad (A.38)$$

it follows that

$$
\begin{aligned}
u_i^*(\mathbf{k}) \, \partial_t u_i(\mathbf{k}) &= -\iota k_j \int u_i^*(\mathbf{k}) \, u_j^*(-\mathbf{k}') \, u_i(\mathbf{k} - \mathbf{k}') d\mathbf{k}' \\
&= -\iota k_j \int u_i^*(\mathbf{k}) \, u_j^*(\mathbf{k}') \, u_i(\mathbf{k} + \mathbf{k}') d\mathbf{k}' \\
&= -\iota k_j \int u_i^*(\mathbf{k}) \, u_j^*(\mathbf{k}') \, u_i^*(-\mathbf{k} - \mathbf{k}') d\mathbf{k}' \\
&= -\iota k_j \int \int \delta(\mathbf{k} + \mathbf{k}' + \mathbf{k}'') \, u_i^*(\mathbf{k}) \, u_j^*(\mathbf{k}') \, u_i^*(\mathbf{k}'') d\mathbf{k}'' d\mathbf{k}', \quad (A.39)
\end{aligned}
$$

using $k_i u_i(\mathbf{k}) = -k_i u_i^*(\mathbf{k}) = 0$ due to the incompressibility, and $u_j(\mathbf{k}') = u_j^*(-\mathbf{k}')$ because $u_j(\mathbf{x})$ is real.

The time derivative of the energy $u_i(\mathbf{k}) u_i^*(\mathbf{k})/2$ at wave vector $\mathbf{k}$ is obtained from (A.39) by adding the complex conjugate and dividing by 2. The Dirac delta function factor shows that a nonzero contribution to the time derivative of the energy at wave vector $\mathbf{k}$ is obtained from wave amplitudes at wave vectors $\mathbf{k}'$ and $\mathbf{k}''$ only if $\mathbf{k} + \mathbf{k}' + \mathbf{k}'' = \mathbf{0}$. A set of three such waves is called a *triad*.

A.8 The 2D case

All the previous results hold also for two-dimensional velocity fields. Summations are just over the two indices 1 and 2. A particular simplification is obtained for the rotation vector rot $\mathbf{u}$ defined in (A.12). Since $u_3 = 0$ and since u_1 and u_2 do not depend on x_3, it can be seen that the two first components vanish, leaving only the third to be different from zero, possibly. For two-dimensional flows (plane flows), the rotation vector therefore reduces to the scalar

$$\omega = \partial_1 u_2 - \partial_2 u_1. \qquad (A.40)$$

Without the force term, the vorticity equation for two-dimensional flows is

$$D_t \omega = \nu \partial_{jj} \omega, \qquad (A.41)$$

which after multiplication by ω and integration becomes the enstrophy equation

$$D_t \int \frac{\omega^2}{2} d\mathbf{x} = \nu \int \omega \, \partial_{jj} \omega \, d\mathbf{x}. \qquad (A.42)$$

Replacing u_i in (A.24) shows that (A.42) can be written as

$$D_t \int \frac{\omega^2}{2}dx = -\nu \int \partial_i \omega \, \partial_i \omega dx. \tag{A.43}$$

Parseval's identity (A.7) (with u_i replaced by ω) applied to (A.43) gives the spectral representation

$$D_t \int \frac{|\omega(\mathbf{k})|^2}{2}d\mathbf{k} = -\nu \int k^2 |\omega(\mathbf{k})|^2 d\mathbf{k}, \tag{A.44}$$

where $k^2 = k_i k_i$. It follows from (A.9) and (A.40) that

$$
\begin{aligned}
\omega(\mathbf{k})\omega^*(\mathbf{k}) &= [k_1 u_2(\mathbf{k}) - k_2 u_1(\mathbf{k})][k_1 u_2^*(\mathbf{k}) - k_2 u_1^*(\mathbf{k})] \\
&= k_1^2 u_2(\mathbf{k})u_2^*(\mathbf{k}) + k_2^2 u_1(\mathbf{k})u_1^*(\mathbf{k}) \\
&\quad - k_1 k_2 [u_2(\mathbf{k})u_1^*(\mathbf{k}) + u_1(\mathbf{k})u_2^*(\mathbf{k})],
\end{aligned} \tag{A.45}
$$

which, by adding the squared spectral incompressibility condition

$$
\begin{aligned}
0 &= [k_1 u_1(\mathbf{k}) + k_2 u_2(\mathbf{k})][k_1 u_1^*(\mathbf{k}) + k_2 u_2^*(\mathbf{k})] \\
&= k_1^2 u_1(\mathbf{k})u_1^*(\mathbf{k}) + k_2^2 u_2(\mathbf{k})u_2^*(\mathbf{k}) \\
&\quad + k_1 k_2 [u_1(\mathbf{k})u_2^*(\mathbf{k}) + u_2(\mathbf{k})u_1^*(\mathbf{k})]
\end{aligned} \tag{A.46}
$$

gives

$$
\begin{aligned}
\omega(\mathbf{k})\omega^*(\mathbf{k}) &= (k_1^2 + k_2^2)[u_1(\mathbf{k})u_1^*(\mathbf{k}) + u_2(\mathbf{k})^* u_2(\mathbf{k})] \\
&= k^2 u_i(\mathbf{k})u_i^*(\mathbf{k}) = k^2 E(\mathbf{k}),
\end{aligned} \tag{A.47}
$$

where $E(\mathbf{k}) = u_i(\mathbf{k})u_i^*(\mathbf{k})/2$. Thus (A.44) becomes

$$D_t \int k^2 E(\mathbf{k})d\mathbf{k} = -2\nu \int k^4 E(\mathbf{k})d\mathbf{k}, \tag{A.48}$$

called the *spectral enstrophy equation*. The *scalar energy spectrum* is defined as

$$E(k) = k \int_0^{2\pi} E(k\cos\theta, k\sin\theta)d\theta, \tag{A.49}$$

and the *scalar enstrophy spectrum* as

$$Z(k) = k^2 E(k). \tag{A.50}$$

It follows from (A.48) that

$$D_t \int_0^\infty Z(k)dk = -2\nu \int_0^\infty k^2 Z(k)dk. \tag{A.51}$$

Without the force term F, the energy Equation (A.25) becomes

$$D_t \int \frac{u_i u_i}{2} d\mathbf{x} = -\nu \int \omega^2 d\mathbf{x}, \tag{A.52}$$

in the 2D case. Parseval's identity (A.7) applied to (A.52) gives the spectral energy equation

$$D_t \int \frac{|u_i(\mathbf{k})|^2}{2} d\mathbf{k} = -\nu \int |\omega(\mathbf{k})|^2 d\mathbf{k}, \tag{A.53}$$

or by (A.47)

$$D_t \int E(\mathbf{k}) d\mathbf{k} = -2\nu \int k^2 E(\mathbf{k}) d\mathbf{k}. \tag{A.54}$$

By (A.49) this equation reduces to

$$D_t \int_0^\infty E(k) dk = -2\nu \int_0^\infty k^2 E(k) dk, \tag{A.55}$$

for the scalar energy spectrum. Thus the scalar enstrophy spectrum and the scalar energy spectrum satisfy the same integral equation.

A.9 Scaling consequence of lognormal assumption

Kolmogorov's and Obukhov's lognormal distribution assumption for the energy dissipation ε_l in (7.1) gives another scaling by l, as well as that based on these authors' crude variance assumption (7.4). This is shown by the following derivation. For any number p we have the mean

$$\langle \varepsilon_l^p \rangle = \langle e^{p \log \varepsilon_l} \rangle$$

$$= \frac{1}{\sqrt{2\pi \operatorname{Var}[\log \varepsilon_l]}} \int_{-\infty}^\infty \exp\left(px - \frac{(x - \langle \log \varepsilon_l \rangle)^2}{2\operatorname{Var}[\log \varepsilon_l]} \right) dx$$

$$= \exp(\langle \log \varepsilon_l \rangle p + \operatorname{Var}[\log \varepsilon_l] p^2 / 2), \tag{A.56}$$

where

$$\operatorname{Var}[\log \varepsilon_l] = \langle (\log \varepsilon_l)^2 \rangle - \langle \log \varepsilon_l \rangle^2 \tag{A.57}$$

is the variance of $\log \varepsilon_l$. Setting $p = 1$ and $p = 2$ in (A.56) gives

$$\log \langle \varepsilon_l \rangle^2 = 2 \langle \log \varepsilon_l \rangle + \operatorname{Var}[\log \varepsilon_l],$$

$$\log \langle \varepsilon_l^2 \rangle = 2 \langle \log \varepsilon_l \rangle + 2\operatorname{Var}[\log \varepsilon_l], \tag{A.58}$$

from which we get

$$\text{Var}[\log \varepsilon_l] = \log \frac{\langle \varepsilon_l^2 \rangle}{\langle \varepsilon_l \rangle^2}. \tag{A.59}$$

Since $\langle \varepsilon_l \rangle = \bar{\varepsilon}$ is independent of l, we have

$$\log \langle \varepsilon_l \rangle = \bar{\varepsilon} - \text{Var}[\log \varepsilon_l]/2. \tag{A.60}$$

Considering now that $\varepsilon_l(\mathbf{x})$ is an average of the random field $F = v(\partial_j u_i + \partial_i u_j)/2$ over the sphere, $|\mathbf{y} - \mathbf{x}| < l$ of center $\mathbf{x}$ and radius l, we have the variance

$$\text{Var}[\varepsilon_l(\mathbf{x})] = V^{-2} \int_{|\mathbf{y}_1 - \mathbf{x}| < l} \int_{|\mathbf{y}_2 - \mathbf{x}| < l} \text{Cov}[F(\mathbf{y}_1), F(\mathbf{y}_2)] d\mathbf{y}_1 d\mathbf{y}_2, \tag{A.61}$$

where $\text{Cov}[F(\mathbf{y}_1), F(\mathbf{y}_2)] = \langle F(\mathbf{y}_1)F(\mathbf{y}_2) \rangle - \langle F(\mathbf{y}_1) \rangle \langle F(\mathbf{y}_2) \rangle$ is the covariance between $F(\mathbf{y}_1)$ and $F(\mathbf{y}_2)$. For simplicity we assume that the random field F is homogeneous. Thus we get

$$\begin{aligned}
\text{Var}[\varepsilon_l] &= \text{Var}[F] v_l^{-2} \int \int \frac{\text{Cov}[F(\mathbf{y}_1), F(\mathbf{y}_2)]}{\text{Var}[F]} d\mathbf{y}_2 d\mathbf{y}_1 \\
&= \text{Var}[F] \left(\frac{R(l)}{l} \right)^3,
\end{aligned} \tag{A.62}$$

where $v_l = \frac{4}{3}\pi l^3$ is the volume of the sphere, and $R(l)$ is defined in the general case of an inhomogeneous field, F, as the radius of a sphere with volume

$$\frac{4}{3}\pi R(l;\mathbf{x})^3 = \int_{|\mathbf{y}-\mathbf{x}|<l} \frac{\text{Cov}[F(\mathbf{x}), F(\mathbf{y})]}{\sqrt{\text{Var}[F(\mathbf{x})]\text{Var}[F(\mathbf{y})]}} d\mathbf{y}. \tag{A.63}$$

The maximal value of $R(l;\mathbf{x})$ with respect to l is called the *correlation radius* at $\mathbf{x}$. Obviously $R(l;\mathbf{x})/l \leq 1$. With the result (A.62) it follows from (A.59) that

$$\text{Var}[\log \varepsilon_l] = \log \left(1 + \frac{\text{Var}[\varepsilon_l]}{\bar{\varepsilon}^2} \right) = \log \left(1 + \frac{R(l)^3}{l^3} V_F^2 \right), \tag{A.64}$$

where $V_F = \sqrt{\text{Var}[F]}/\bar{\varepsilon}$ is the coefficient of variation of $F(\mathbf{x})$. Substituting (A.64) into (A.56) and thereafter using (A.62) we get

$$\begin{aligned}
\langle \varepsilon_l^p \rangle &= \exp[p\bar{\varepsilon} + (p^2 - p)\text{Var}[\log \varepsilon_l]/2] \\
&= \exp \left\{ p\bar{\varepsilon} + \frac{1}{2}\log \left[1 + \frac{R(l)^3}{l^3} V_F^2 \right] (p^2 - p) \right\}. \tag{A.65}
\end{aligned}$$

With $R(l)$ unknown in detail, we can simplify the ratio $R(l)/l$ to the upper bound $\min\{1, R_{\max}/l\}$. Thus

$$
\langle \varepsilon_l^p \rangle \leq
\begin{cases}
\mathcal{O}(1) & \text{for } l \leq R_{\max} \\
\exp\left\{ \dfrac{1}{2} \dfrac{R_{\max}^3}{l^3} V_F^2 (p^2 - p) \right\} & \text{for } l > R_{\max}
\end{cases}.
\tag{A.66}
$$

References

André, J. C., & Lesieur, M. 1977. Influence of helicity on high Reynolds number isotropic turbulence. *J. Fluid Mech.*, **81**, 187–207.

Anselmet, F., Gagne, Y., Hopfinger, E. J., & Antonia, R. A. 1984. High-order velocity structure functions in turbulent shear flow. *J. Fluid Mech.*, **140**, 63–89.

Aurell, E., Boffetta, G., Crisanti, A., *et al.* 1994. Statistical mechanics of shell models for two-dimensional turbulence. *Phys. Rev. E*, **50**, 4705–15.

Aurell, E., Boffetta, G., Crisanti, A., Paladin, G., & Vulpiani, A. 1996a. Growth of noninfinitesimal perturbations in turbulence. *Phys. Rev. Lett.*, **77**, 1262–5.

Aurell, E., Boffetta, G., Crisanti, A., Paladin, G., & Vulpiani, A. 1996b. Predictability in systems with many characteristic times: The case of turbulence. *Phys. Rev. E*, **53**, 2337–49.

Biferale, L., Lambert, A., Lima, R., & Paladin, G. 1995. Transition to chaos in a shell model of turbulence. *Physica D*, **80**, 105–19.

Biferale, L., Pierotti, D., & Toschi, F. 1998. Helicity transfer in turbulent models. *Phys. Rev. E*, **57**, R2515–8.

Boer, G. J., & Shepherd, T. G. 1983. Large-scale two-dimensional turbulence in the atmosphere. *J. Atmos. Sci.*, **40**, 164–84.

Bohr, T., Jensen, M., Paladin, G., & Vulpiani, A. 1998. *Dynamical Systems Approach to Turbulence*. Cambridge, Cambridge University Press.

Borue, V., & Orszag, S. A. 1997. Spectra in helical three-dimensional homogeneous isotropic turbulence. *Phys. Rev. E*, **55**, 7005–9.

Brissaud, A., Frisch, U., Leorat, J., Lesieur, M., & Mazure, A. 1973. Helicity cascade in fully developed isotropic turbulence. *Phys. Fluids*, **16**, 1366–7.

Charney, J. G. 1971. Geostropic turbulence. *J. Atmos. Sci.*, **28**, 1087–95.

Charney, J. G., & Stern, M. E. 1962. On the stability of internal baroclinic jets in a rotating atmosphere. *J. Atmos. Sci.*, **19**, 159–72.

Chkhetiani, O. G. 1996. On the third moments in helical turbulence. *JETP Lett.*, **63**, 808–12.

Clay Mathematics Institute. 2000. *Seven Problems of the Millenium*. www.claymath.org/prizeproblems.

Comte-Bellot, G., & Corrsin, S. 1971. Simple Eulerian time correlation of full- and narrow-band velocity signals in grid-generated, 'isotropic' turbulence. *J. Fluid Mech.*, **48**, 273–337.

Crisanti, A., Jensen, M. H., Vulpiani, A., & Paladin, G. 1993a. Intermittency and predictability in turbulence. *Phys. Rev. Lett.*, **70**, 166–9.

Crisanti, A., Vulpiani, A., Jensen, M. H., & Paladin, G. 1993b. Predictability and the butterfly effect in turbulent flows: A shell model study. *Int. J. Bifurcation Chaos*, **3**, 1581–5.

Desnyansky, V. N., & Novikov, E. A. 1974. The evolution of turbulence spectra to the similarity regime. *Izv. Akad. Nauk SSSR, Fiz. Atmos. Okeana*, **10**, 127–36.

Ditlevsen, P. D. 1996. Temporal intermittency and cascades in shell models of turbulence. *Phys. Rev. E.*, **54**, 985–8.

Ditlevsen, P. D. 1997. Cascades of energy and helicity in the GOY shell model of turbulence. *Phys. Fluids*, **9**, 1482–4.

Ditlevsen, P. D. 2000. Symmetries, invariants, and cascades in a shell model of turbulence. *Phys. Rev. E*, **62**, 484–9.

Ditlevsen, P. D., & Giuliani, P. 2000. Anomalous scaling in a shell model of helical turbulence. *Physica A*, **280**, 69–74.

Ditlevsen, P. D., & Giulani, P. 2001a. Cascades in helical turbulence. *Phys. Rev. E.*, **63**, 036304.

Ditlevsen, P. D., & Giuliani, P. 2001b. Dissipation in helical turbulence. *Phys. Fluids*, **13**, 3508–9.

Ditlevsen, P. D., & Mogensen, I. A. 1996. Cascades and statistical equilibrium in shell models of turbulence. *Phys. Rev. E*, **53**, 4785–93.

Ditlevsen, P. D., Jensen, M. H., & Olesen, P. 2004. Scaling and the prediction of energy spectra in decaying hydrodynamic turbulence. *Physica A*, **342**, 471–8.

Frick, P., & Sokoloff, D. 1998. Cascade and dynamo action in a shell model of magnetohydrodynamic turbulence. *Phys. Rev. E*, **57**, 4155–64.

Frisch, U. 1995. *Turbulence: The Legacy of A. N. Kolmogorov*. Cambridge University Press, Cambridge.

Frisch, U., & Parisi, G. 1985. On the singular structure of fully developed turbulence. Pages 84–87 of: Ghil, M., Benzi, R., & Parisi, G. (eds.), *Turbulence and Predictability in Geophysical Fluid Dynamics and Climate Dynamics*. Amsterdam/New York/Oxford/Tokyo: North-Holland Publ. Co.

Gibson, J. K., Källberg, P., Uppala, S., *et al.* 1997. *ECMWF re-analysis report series, vol. 1. ERA Description*. European Center for Medium Range Weather Forecast, Reading, UK, 72pp.

Gledzer, E. B. 1973. System of hydrodynamic type admitting two quadratic integrals of motion. *Sov. Phys. Dokl. SSSR*, **18**, 216–7.

Goldstein, H. 1980. *Classical Mechanics*. 2nd edition. Reading, MA, Addison-Wesley.

Grassberger, P., & Procaccia, I. 1983. Measuring the strangeness of strange attractors. *Physica D*, **9**, 189–208.

Grigoriu, M., O., Ditlevsen, P. D., & Arwade, S. R. 2003. A Monte Carlo simulation model for stationary non-Gaussian processes. *Probabilistic Engineering Mechanics*, **18**, 87–95.

Grossmann, S., & Lohse, D. 1994. Universality in fully-developed turbulence. *Phys. Rev. E*, **50**, 2784–9.

Gurbatov, S. N., Simdyankin, S. I., Aurell, E., Frisch, U., & Tóth, G. 1997. On the decay of Burgers turbulence. *J. Fluid Mech.*, **344**, 339–74.

Holton, J. R. 1992. *An Introduction to Dynamic Meteorology*. Third edition. San Diego, Academic Press.

Jensen, M. H., Paladin, G., & Vulpiani, A. 1991. Intermittency in a cascade model for three-dimensional turbulence. *Phys. Rev. A*, **43**, 798–805.

Jensen, M. H., Paladin, G., & Vulpiani, A. 1992. Shell model for turbulent advection of passive-scaler fields. *Phys. Rev. A*, **45**, 7214–21.

Kadanoff, L., Lohse, D., Wang, J., & Benzi, R. 1995. Scaling and dissipation in the GOY shell model. *Phys. Fluids*, **7**, 617–29.

Kadanoff, L., Lohse, D., & Schörghofer, N. 1997. Scaling and linear response in the GOY turbulence model. *Physica D*, **100**, 165–86.

Kang, H. S., & Meneveau, C. 2001. Passive scalar anisotropy in a heated turbulent wake: New observations and implications for large-eddy simulations. *J. Fluid Mech.*, **442**, 161–70.

Kaplan, J. L., & Yorke, J. A. 1979. Preturbulence-regime observed in a fluid-flow model of Lorenz. *Commun. Math. Phys.*, **67**, 93–108.

Kerr, R., & Biferale, L. 1995. On the role of inviscid invariants in shell models of turbulence. *Phys. Rev. E*, **52**, 6113–22.

Kolmogorov, A. N. 1941a. Dissipation of energy in locally isotropic turbulence. *C. R. (Dokl.) Acad. Sci. SSSR*, **32**, 16–18.

Kolmogorov, A. N. 1941b. Local structure of turbulence in an incompressible liquid for very large Reynolds numbers. *C. R. (Dokl.) Acad. Sci. SSSR*, **30**, 299.

Kolmogorov, A. N. 1941c. On degeneration (decay) of isotropic turbulence in an incompressible viscous liquid. *C. R. (Dokl.) Acad. Sci. SSSR*, **31**, 538–540.

Kolmogorov, A. N. 1962. A refinement of previous hypothesis concerning the local structure of turbulence in a viscous incompressible fluid at high Reynolds number. *J. Fluid Mech.*, **5**, 82–5.

Kraichnan, R. H. 1967. Inertial ranges in two-dimensional turbulence. *Phys. Fluids*, **10**, 1417–23.

Kraichnan, R. H. 1971. Inertial-range transfer in 2-dimensional and 3-dimensional turbulence. *J. Fluid Mech.*, **47**, 525–35.

Kraichnan, R. H., & Montgomery, D. 1980. Two-dimensional turbulence. *Rep. Prog. Phys.*, **43**, 547–619.

Landau, L. D., & Lifshitz, E. M. 1987. *Fluid Mechanics*, 2nd edition. Oxford, Pergamon Press.

Leray, J. 1934. Sur le mouvement d'un liquide visquex emplissement l'espace. *Acta Math. J.*, **63**, 193–248.

Lesieur, M. 1997. *Turbulence in Fluids*, 3rd Revised and enlarged edition. Dordrecht, Kluwer Academic Publishers.

Lindborg, E. 1999. Can the atmospheric kinetic energy spectrum be explained by two-dimensional turbulence? *J. Fluid Mech.*, **388**, 259–88.

Lorenz, E. N. 1963. Deterministic nonperiodic flow. *J. Atmos. Sci.*, **20**, 130–41.

Lorenz, E. N. 1969. The predictability of a flow which possesses many scales of motion. *Tellus*, **21**, 289–307.

Lorenz, E. N. 1972. Low order models representing realizations of turbulence. *J. Fluid Mech.*, **55**, 545–63.

L'vov, V. S., Podivilov, E., & Procaccia, I. 1997. Exact result for the 3rd order correlations of velocity in turbulence with helicity. chao-dyn/9705016.

L'vov, V. S., Podivilov, E., Pomyalov, A., Procaccia, I., & Vandembroueq, D. 1998. Improved shell model of turbulence. *Phys. Rev. E*, **58**, 1811–22.

Mandelbrot, B. 1977. *Fractals: Form, Chance and Dimension*. San Francisco, Freeman & Co.

Mohamed, M. S., & LaRue, J. C. 1990. The decay power law in grid-generated turbulence. *J. Fluid Mech.*, **219**, 195–214.

Monin, A. S., & Yaglom, A. M. 1981. *Statistical Fluid Mechanics: Mechanics of Turbulence, Volumes 1+2*. Cambridge, MA, The MIT Press.

Nastrom, G. D., & Gage, K. S. 1985. A climatology of atmospheric wavenumber spectra observed by commercial aircraft. *J. Atmos. Sci.*, **42**, 950–60.

Obukhov, A. M. 1962. Some specific features of atmospheric turbulence. *J. Fluid Mech.*, **13**, 77–81.

Obukhov, A. M. 1971. On some general characteristics of the equations of the dynamics of the atmosphere. *Izv. Akad. Nauk SSSR, Fiz. Atmos. Okeana*, **7**, 695–704.

Olesen, P. 1997. Inverse cascades and primordial magnetic fields. *Phys. Lett. B*, **398**, 321–25.

Olla, P. 1998. Three applications of scaling to inhomogeneous, anisotropic turbulence. *Phys. Rev. E*, **57**, 2824–31.

Pedlosky, J. 1987. *Geophysical Fluid Dynamics*, 2nd edition. New York, Springer-Verlag.

Pope, S. B. 2000. *Turbulent Flows*. Cambridge, Cambridge University Press.

Richardson, L. F. 1922. *Weather Prediction by Numerical Process*. Cambridge, Cambridge University Press.

Ruelle, D. 1979. Microscopic fluctuations and turbulence. *Phys. Lett. A*, **72**, 81–2.

Saltzman, B. 1962. Finite amplitude free convection as an initial value problem. *J. Atmos. Sci.*, **19**, 329–41.

Shannon, C. E. 1948. A mathematical theory of communication. *Bell System Technical Journal*, **27**, 379–423.

She, Z. S., Aurell, E., & Frisch, U. 1992. The inviscid Burgers equation with initial conditions of Brownian type. *Comm. Math. Phys.*, **148**, 623–641.

Straus, D. M., & Ditlevsen, P. 1999. Two-dimensional turbulence properties of the ECMWF reanalysis. *Tellus*, **51A**, 749–72.

Vergassola, M., Dubrulle, B., Frisch, U., & Noullez, A. 1994. Burger's equation, Devil's staircases and the mass distribution for large-scale structures. *Astron. Astrophys.*, **289**, 325–56.

von Kármán, T., & Howarth, L. 1938. On the statistical theory of isotropic turbulence. *Proc. R. Soc. London*, **A 164**, 192–215.

Waleffe, F. 1992. The nature of triad interactions in homogeneous turbulence. *Phys. Fluids A-Fluid Dynamics*, **4**, 350–63.

Wiin-Nielsen, A. 1967. On annual variation and spectral distribution of atmospheric energy. *Tellus*, **19**, 540–59.

Wiin-Nielsen, A. 1972. A study of power laws in the atmospheric kinetic energy spectrum using spherical harmonic functions. *Meteor. Ann.*, **6**, 107–24.

Yamada, M., & Okhitani, K. 1988a. Lyapunov spectrum of a model of two-dimensional turbulence. *Phys. Rev. Lett.*, **60**, 983–6.

Yamada, M., & Okhitani, K. 1988b. The inertial subrange and non-positive Lyapunov exponents in fully-developed turbulence. *Progr. Theo. Phys.*, **79**, 1265–68.

Index

Printed in the United States
by Bookmasters

Printed in the United States
By Bookmasters